Springer Undergraduate Mathematics Series

More information about this series at http://www.springer.com/series/3423

Roozbeh Hazrat

Mathematica®:
A Problem-Centered
Approach

Second Edition

 Springer

Roozbeh Hazrat
Centre for Research in Mathematics
Western Sydney University
Sydney
Australia

ISSN 1615-2085   ISSN 2197-4144   (electronic)
Springer Undergraduate Mathematics Series
ISBN 978-3-319-27584-0   ISBN 978-3-319-27585-7   (eBook)
DOI 10.1007/978-3-319-27585-7

Library of Congress Control Number: 2015958536

Mathematics Subject Classification (2010): 68-01, 68W30

Printed on acid-free paper

This Springer imprint is published by SpringerNature
The registered company is Springer International Publishing AG Switzerland

به پدرم و به مادرم

و به شهره و آزاده

Preface to the second edition

Since the publication of the book, I have had the opportunity to run further courses at Queen's University Belfast, Western Sydney University and training courses at the Reserve Bank of Australia. I also received valuable input from the readers in the last 6 years. The second edition takes all of this into account. We have updated the materials, added many more examples and problems and have added a new chapter "Projects" which demonstrates how to use *Mathematica* to explore mathematics. Some of these projects were given as assignments to the students in the courses mentioned above.

Most of the work on the second edition was done during the ferry rides Manly-Circular Quay-Parramatta to work. I would like to thank Sydney Ferries for running such a smooth service.

Roozbeh Hazrat
r.hazrat@uws.edu.au
Sydney, October 2015

Preface to the first edition

This book grew out of a course I gave at Queen's University Belfast during the period 2004 to 2009. Although there are many books already written on how to use Wolfram *Mathematica*®, I noticed that they fall into two categories: either they provide an explanation of how to use the commands, in the style of: enter the command, push the button and see the result; or they study some problems and write several-paragraph codes in *Mathematica*. The books in the first category did not inspire me (or my imagination) and those in the second category were too difficult to understand and were not suitable for learning (or teaching) *Mathematica*'s abilities to do programming and solve problems.

I could not find a book that I could follow to teach this module. In class one cannot go on forever showing students just how commands in *Mathematica* work; on the other hand it is very difficult to follow raw codes having more than five lines (especially as *Mathematica*'s style of programming provides a condensed code). Thus this book.

This book promotes *Mathematica*'s style of programming. I tried to show when we adopt this approach, then how naturally one can solve (nice) problems with (*Mathematica*) style.

Here is an example: Does the Euler formula $n^2 + n + 41$ produce prime numbers for $n = 1$ to 39?

```
(#^2 + # + 41) & /@ Range[39] ∈ Primes
```

```
True
```

In another problem, for example, we try to show how one can effectively use pattern matching to check that for two matrices A and B, $(ABA^{-1})^5 =$

AB^5A^{-1}. One only needs to introduce the fact that $A^{-1}A = 1$ and then *Mathematica* will verify the result by cancelling the inverse elements instead of by direct calculation.

Although the meaning of the code above may not be clear yet, the reader will observe as we proceed how the codes start to make sense, as if this is the most natural way to approach the problems. (People who approach problems with a procedural style of programming (such as C++) will experience that this style replaces their way of thinking!) We have tried to let the reader learn from the codes and avoid long and exhausting explanations, as the codes will speak for themselves. Also we have tried to show that in *Mathematica* (as in the real world) there are many ways to approach a problem and solve it. We have tried to inspire the imagination!

Someone once rightly said that the *Mathematica* programming language is rather like a "Swiss army knife" containing a vast array of features. *Mathematica* provides us with powerful mathematical functions. Along with this, one can freely mix different styles of programming, functional, list-based and procedural, to achieve a lot. This mélange of programming styles is what we promote in this volume.

Thus this book could be considered for a course in *Mathematica*, or for self study. It mainly concentrates on programming and problem solving in *Mathematica*. I have mostly chosen problems having something to do with natural numbers as they do not need any particular background. Many of these problems were taken from or inspired by those collected in [4].

I would like to thank Ilan Vardi for answering my emails and Brian McMaster and Judith Millar for polishing the English.

Naoko Morita encouraged me to turn my notes into this book. I thank her for this and for always smiling and having a little Geschichte zu erzählen.

Roozbeh Hazrat

r.hazrat@qub.ac.uk

Belfast, October 2009

How to use this book

Each chapter of the book starts with a description of a new topic and some basic examples. We will then demonstrate the use of new commands in several problems and their solutions. We have tried to let the reader learn from the codes and avoided long and exhausting explanations, as the codes will speak for themselves.

There are three different categories of problems, indicated by different frames:

▬▬ Problem 0.1

These problems are the essential parts of the text where new commands are introduced to solve the problem and where it is demonstrated how the commands are used in Wolfram *Mathematica*®. These problems should not be skipped.

⟹ SOLUTION.

▬▬ Problem 0.2

These problems further demonstrate how one can use commands already introduced to tackle different situations. The readers are encouraged to try out these problems by themselves first and then look at the solution.

⟹ SOLUTION.

_____ Problem 0.3

These are more challenging problems that could be skipped in the first reading.

$\Longrightarrow$ SOLUTION.

Most commands in *Mathematica* have several extensions. When we introduce a command, this is just the tip of the iceberg. In TIPS, we give further indications about the commands, or new commands to explore. Once one has enough competence, then it is quite easy to learn further features and abilities of a command by using the *Mathematica* Help and the examples available there.

Suggested Course Outlines

The materials of this book have been used for a one semester long course at an undergraduate level as well as one and two day training courses for PhD students and experienced programmers.

The course could be a combination of lectures and lab work. The materials are presented to the students in the class where the instructor types the codes and explains the structures. There will then be lab sessions where students work on the problems and exercises of the book by themselves. They will then work on Projects A and B of Chapter 17 and will submit them for possible credits. Each lecture and lab session lasts 90 minutes.

Sessions	Contents
Lecture 1	Chapters: 1, 2, 3.1
Lecture 2	Chapters: recall 3.1, 3.2, 4.1, 4.2
Lab session 1	Exercises of Chapters 1 to 4.2
Lecture 3	Chapters: 4.3, 5, 6
Lab session 2	Exercises of Chapters 4 to 6
Lecture 4	Chapters 7, 8
Lab session 3	Exercises of Chapters 7 and 8
Lecture 5	Chapters 9, 10, 11.1
Lab session 4	Exercises of Chapters 9 to 11 and Projects A
Lecture 6	Chapters 11.2, 11.3, 12
Lab session 5	Exercises of Chapters 11 to 12 and Projects A
Lecture 7	Chapters 13, 14, 15
Lab session 6	Exercises of Chapters 11 to 12 and Projects A
Lecture 8	Chapter 16.1 or 16.2 or 16.3.
Lab session 7	Chapter 17, Projects A
Lab sessions 8 to 12	Chapter 17, Projects B

The Mathematica philosophy

In the beginning is the expression. Wolfram *Mathematica*® transforms the expression dictated by the rules to another expression. And that's how a new idea comes into the world.

The rules that will be used frequently, can be given a name (we call them functions)

```
r[x_]:=1+x^2

r[arrow]
1+arrow^2

r[5]
26
```

And the transformation could take place immediately or with a delay

```
{x,x}/.x->Random[]
{0.0474307, 0.0474307}

{x,x}/.x:>Random[]
{0.368461, 0.588353}
```

The most powerful transformations are those which look for a certain pattern in an expression and morph that to a new pattern.

```
(a + b)^c /. (x_ + y_)^z_ -> (x^z + y^z)
a^c + b^c
```

And one of the most important expressions that we will work with is a list. As the name implies, this is just a collection of elements (collection of other expressions). We then apply transformations to each element of the list:

```
x^ {1, 5, 7}
{x, x^5, x^7}
```

Any expression is considered as having a "head" followed by several arguments, head[arg1,arg2,...]. And one of the transformations which provide a capable tool is to replace the head of an expression by another head!

```
Plus @@ {a,b,c}
a+b+c
```

```
Power @@ (x+y)
x^y
```

Putting these concepts together creates a powerful way to solve problems.

In the chapters of this book, we decipher these concepts.

Contents

1
Introduction

This first chapter gives an introduction to using *Mathematica* "out of the box", demonstrating how to use ready-made commands, performing basic operations, building up computations, creating professional and magnificent two- and three-dimensional graphics and creating dynamic and interactive presentations.

We go through academic life to enhance our problem solving techniques. These days computational software is an indispensable tool for solving problems. In this book we focus on the art of problem solving, using the software Wolfram *Mathematica*®, which is one of the most powerful softwares available.

There are at least two features of *Mathematica* which make it an excellent software. It provides many powerful, easy to use, functions which allows one to calculate, investigate and analyse problems. These functions, which have very consistent structures, can be used in *Mathematica*'s editor (front-end).

The other feature is the way *Mathematica* allows one to put these functions together and to build up a program to solve problems.

In this chapter we will look at the available functions in *Mathematica*. We will give an overview of what *Mathematica* can do "out of the box".

1.1 Basic calculations

Mathematica handles all sort of numerical calculations, both exact computations and approximations. If we would like to calculate 2^{3^4} or $\log(\cos(e^{\sqrt{4.7}}))$, we only need to enter them correctly into *Mathematica*.

© Springer International Publishing Switzerland 2015
R. Hazrat, *Mathematica®: A Problem-Centered Approach*, Springer Undergraduate
Mathematics Series, DOI 10.1007/978-3-319-27585-7_1

```
2^3^4
2417851639229258349412352
```

```
Log[Cos[E^(Sqrt[4.7])]]
-0.255189 + 3.14159 I
```

One uses square brackets [ ] to pass data into functions, such as, `Log[4.7]` and round brackets ( ) to group the expressions together.

```
(2^3)^4
4096
```

```
2^(3^4)
2417851639229258349412352
```

```
Log[4.7]
1.54756
```

One can use *Mathematica*'s Basic Math Assistant (from Palettes) to enter mathematical expressions as is:

$$\sqrt[3]{e^\pi + \log\left(\frac{23}{\sin\left(\frac{\pi}{6}\right)}\right)}$$

$$\sqrt[3]{e^\pi + \log(46)}$$

Mathematica didn't calculate and approximate the expression above. We can ask for the numerical value of this expression by using the built-in function N.

$$N\left[\sqrt[3]{e^\pi + \log\left(\frac{23}{\sin\left(\frac{\pi}{6}\right)}\right)}\right]$$

2.99886

For more precision we can use N as follows:

$$N\left[\sqrt[3]{e^\pi + \log\left(\frac{23}{\sin\left(\frac{\pi}{6}\right)}\right)}, 50\right]$$

2.9988637930384753371494836101827755196014972898262

One can get a short description of what the built-in functions do as follows:

```
? N

N[expr] gives the numerical value of expr.
N[expr,n] attempts to give a result with n-digit precision
```

In fact, for some involved computations, this precision needs to be used, otherwise *Mathematica* might give a wrong result, as the following example shows.

```
Sin[(E + Pi)^100]
Sin[(E + Pi])^100]

N[Sin[(E + Pi)^100]]
-0.747485

N[Sin[(E + Pi)^100.]]
0.882832

N[Sin[(E + Pi)^100.], 40]
0.882832

N[Sin[(E + Pi)^100], 30]
```

During evaluation of In[67]:= N::meprec: Internal precision
limit $MaxExtraPrecision = 50.' reached while evaluating
Sin[(E+Pi])^100]. >>

0.799751375979095575644

Here is a rather more advanced way to get *Mathematica* to calculate with
more precision. We can ask *Mathematica* to set the precision up to 500 significant digits. The command Block ensures that this change of setting only
applies to the code in that block (see Section 11.1).

```
?$MaxExtraPrecision
```

$MaxExtraPrecision gives the maximum number of extra digits of
precision to be used in functions such as N. >>

```
Block[
{$MaxExtraPrecision = 500},
  N[Sin[(E + Pi)^100], 300]
]
```

0.7997513759790955756444728813732518997662614556734125676073
921019476912401368160187306781505824759366709891236181814372
546109449995738739978325777898970629951717472201054190170248
914583373319286665852874714602388172056308177924956640136927
360031508204693327449015265171180089289328002848076083037364
51268937113801627373002338898735699558404324208681166259134
072835670505777174604982434774860870275355

1.2 Graphics

Mathematica provides many graphical built-in functions. Here are two interesting ones. CurrentImage captures a photo via the computer's camera. Then the command EdgeDetect recognises the edges of the picture and creates a black and white image. These might not be very impressive as a smartphone app can do all this. However, an app functionality ends here whereas, with *Mathematica*, things start from here.

```
x=EdgeDetect[CurrentImage[]]
```

Once we assign the image to the variable x we are able to manipulate the image with the available functions. Here we rotate the image using the function Rotate.

```
Table[Rotate[x, t], {t, 0, 360, 60}]
```

Here is another amusing example. We create a loop, taking 8 consecutive photos and then apply different effects to them.

```
x8 = Table[Pause[0.5]; CurrentImage[], {8}]
```

The command `Table` creates a *loop*, repeating the functions inside it eight times. We will study this function in Chapter 4.

We proceed by assigning different effects to these 8 shots. We use the command `Inner` which will be explored in Chapter 8.

```
effects ={"Charcoal","Embossing","OilPainting","Posterization",
   "Solarization","MotionBlur","ColorBlindness","Comics"};

Inner[ImageEffect, x8, effects, List]
```

It is very easy to plot the graph of functions with *Mathematica*. If we are interested in visualising a function $f(x)$, all we need to do is to use the command `Plot` and give *Mathematica* the range where we would like to plot the graph. For example, the graph of $f(x) = \sin(x)/x$ for $-40 \leq x \leq 40$ is created as follows:

```
Plot[Sin[x]/x, {x, -40, 40}, PlotRange -> All]
```

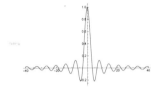

Later in Chapter 15 we analyse the function $\sin(x)/x$, calculating its limit as x approaches 0.

It is equally easy to plot several functions together. One puts these functions inside a *list*, { }, and uses the function `Plot`.

```
Plot[{Sin[Cos[x]], Cos[Sin[x]]}, {x, -2 Pi, 2 Pi},
    PlotLegends -> LineLegend["Expressions"]]
```

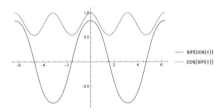

As can be seen from the code above, one can use many optional commands to "decorate" the output.

```
Plot[{Sin[Cos[x]], Cos[Sin[x]]}, {x, -2 Pi, 2 Pi},
    Ticks -> {{-2 Pi, -3 Pi/2, -Pi, -Pi/2, 0, Pi/2,
    {Pi, Rotate[Text[Style["point", Italic, Bold, Red]],90 Degree]},
        3 Pi/2, 2 Pi}}, PlotLegends -> LineLegend["Expressions"],
    Background -> Lighter[Gray, 0.5]]
```

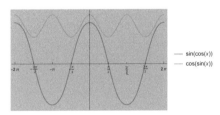

If a function is described as a parametric equation, i.e., the x and y depend on other parameters, one can use `ParametricPlot` to plot the graph of the function. For example, consider the complex function,

$$g(t) = e^{-2it} + \frac{1}{2}e^{5it} + \frac{1}{5}e^{19it},$$

and the graph where x and y are the real and imaginary part of $g(t)$ for different values of t. We can plot this graph in *Mathematica* with ease. First we define the function $g(t)$ in *Mathematica*. Defining functions is the theme of Chapter 3.

```
g[t_] := Exp[-2 I t] + 1/2 Exp[5 I t] + 1/5 Exp[19 I t]
```

The commands Re and Im return the real and imaginary part of a complex number.

```
ParametricPlot[{Re[g[t]], Im[g[t]]}, {t, 0, 2 Pi}]
```

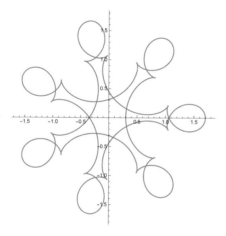

Manipulate is a powerful command which will be used to show the changes of a function dynamically. One can create interactive outputs using this command. We will see several applications of this command throughout this book, not only in the setting of graphics, but in many other instances.

```
Manipulate[expr,{u, umin, umax}] generates a version of expr
with controls added to allow interactive manipulation of the
value of u.
```

```
Manipulate[ParametricPlot[{Re[g[t]], Im[g[t]]}, {t, 0, x}],
    {x, 0.1, 2 Pi}]
```

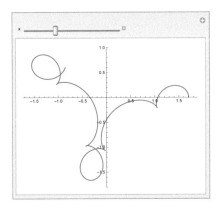

Another example of `Manipulate` to observe the effect of different meshing of the graph is presented in the code below.

```
Manipulate[Plot[{x + 4 Sin[x ^2], x - Cos[x]}, {x, 0, 2 Pi},
    Mesh -> i, MeshStyle -> Red], {i, 1, 100, 1}]
```

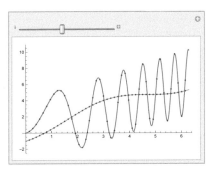

Here is the equation of the so-called bouncing wagon

$$2y^3 + y^2 - y^5 = x^4 - 2x^3 + x^2.$$

The command `ContourPlot` will find all the points (x, y) which satisfy this equation and thus justifies why the function is called the bouncing wagon.

```
ContourPlot[2 y^3 + y^2 - y^5 == x^4 - 2 x^3 + x^2,
    {x, -5, 5}, {y, -5, 5}]
```

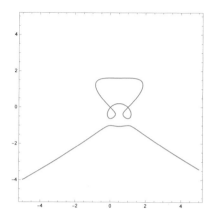

Mathematica can easily plot three-dimensional graphics. Similar to two-dimensional graphs, one uses `Plot3D` to plot a function of the form $f(x, y)$. In the following example we plot the function $\cos(x^2 + y^2)$ in the region $-\pi \le x \le \pi$ and $-\pi \le y \le \pi$.

```
Plot3D[Cos[x^2 + y^2], {x, -Pi, Pi}, {y, -Pi, Pi}]
```

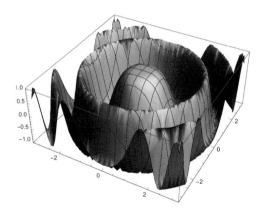

Mathematica allows us to enhance the graph, change many aspects of it, etc. One can, for example, ask *Mathematica* to create a finer, more detailed plot using `PlotPoints`.

```
Plot3D[Cos[x^2 + y^2], {x, -Pi, Pi}, {y, -Pi, Pi},
  PlotPoints -> 50]
```

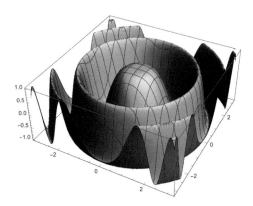

Next we ask *Mathematica* to rotate the graph of $\sin(x)$ around the y axis to create a three-dimensional graph. Can you imagine what the result looks like?

```
RevolutionPlot3D[Sin[x], {x, 0, 4 Pi}]
```

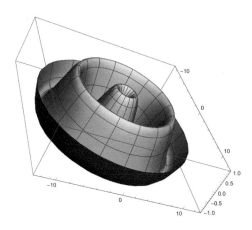

Next we generate a graph of a parametric function $(\sin(3t), \cos(4t))$ and then ask *Mathematica* to rotate this function.

```
ParametricPlot[{Sin[3 t], Cos[4 t]}, {t, 0, Pi}]
```

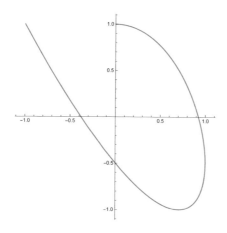

RevolutionPlot3D[{Sin[3 t], Cos[4 t]}, {t, 0, Pi}]

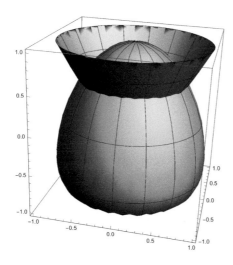

We now consider the function $\sin(x^2)\cos(y^2)$ and plot its contour by using
ContourPlot.

ContourPlot[Sin[x^2] Cos[y^2], {x, -Pi, Pi}, {y, -Pi, Pi}]

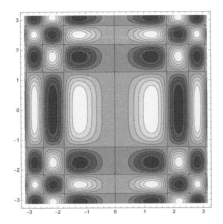

We then plot the function and its contour using `Manipulate`, observing how the areas change as the plot develops.

```
Manipulate[
 GraphicsRow[{
   Plot3D[Sin[x^2] Cos[y^2],{x, -Pi, -Pi+ i},{y, -Pi, -Pi+ i}],
   ContourPlot[
    Sin[x^2] Cos[y^2], {x, -Pi, -Pi+ i}, {y, -Pi, -Pi+ i}]}],
   {i, 0.1, 2 Pi}]
```

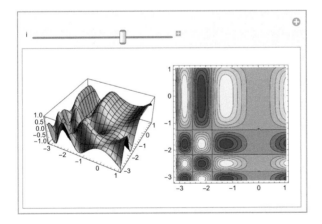

We will study the graphical capabilities of *Mathematica* in Chapter 14.

1.3 Handling data

Mathematica can handle data easily and efficiently. One can generate data, arrange them in different formats and export them in different types. Below we give an example of data coming from points generated by an equation. We then export this data into a spreadsheet.

The equation we are dealing with is $\sin(x^2) - \cos(x)$. We plot this equation, shifting it up by 3 steps. In order to do this, we create a list containing $\sin(x^2) - \cos(x), \sin(x^2) - \cos(x) + 3$ and $\sin(x^2) - \cos(x) + 6$. Here is the result.

```
Plot[{Sin[x^2] - Cos[x], Sin[x^2] - Cos[x] + 3,
   Sin[x^2] - Cos[x] + 6}, {x, 0, Pi}]
```

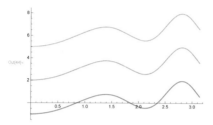

In order to generate (interesting) data using these equations, we randomly change the output values by `Random[]`.

```
xSample :=
 Table[Random[] + Sin[x^2] - Cos[x] + i,
   {i, 0, 6, 3}, {x, 0, Pi, 0.1}]
```

In the above code, we have used the function `Table` which as we mentioned, creates a loop, repeating the function given to it, by running i from 0 to 6 and and x from 0 to π. Let us look at `xSample` and the different arrangements of the data it contains.

```
xSample
{{-0.322704, -0.16157, -0.648524, -0.484712, -0.314378,
  0.134655, 0.510873, -0.238919, 0.389404, 1.02489, 0.49493, 0.961443,
  1.58049, 1.71249, 0.88916, 0.819947, 1.52838, 1.37173, 0.767405,
  -0.0785505, 0.394785, 0.376431, -0.401597, 0.176351, 0.295896,
  0.770743, 2.02606, 2.7163, 2.5529, 2.05827, 2.12839, 1.72677},
 {2.12186, 2.33073, 2.59245, 2.56692, 2.40873, 2.69847, 2.92555, 3.02555,
  3.12114, 3.43739, 4.06132, 3.75216, 4.11431, 4.2339, 4.51363, 4.62963,
  4.00562, 3.88351, 3.17712, 3.82702, 3.47563, 2.81799, 2.91852, 2.8719,
  3.93217, 4.72, 5.1052, 5.36023, 5.46607, 5.44371, 4.79286, 4.10623},
 {5.30329, 5.30367, 5.69052, 5.15566, 6.05634, 6.15, 6.39915, 5.8046,
  6.29151, 6.37714, 7.12522, 6.62564, 7.20384, 6.73209, 7.25741, 6.80756,
  7.45887, 6.43244, 6.84176, 6.36102, 6.01572, 5.98147, 5.91868, 6.02645,
  6.29083, 6.91054, 8.00729, 7.92604, 8.17712, 8.18272, 8.22135, 6.89298}}
```

Mathematica provides several commands to arrange and present the lists (data) on the screen. Try these two: `TableForm` and `TableView`.

```
Transpose[xSample] // TableView
```

	1	2	3	4	5
1	-.016	2.456	5.099		
2	-.057	2.854	5.113		
3	-.090	2.369	5.578		
4	-.072	2.920	5.744		
5	-.063	3.2159	5.249		
6	-.048	2.6046	6.343		
7	0.082	3.365	6.442		
8	-.018	3.341	5.718		
9	0.567	3.0776	6.077		
10	0.642	3.153	6.473		
11	0.720	4.223	6.897		
12	0.896	3.533	7.209		
13	1.108	3.716	7.0955		
14	1.330	3.849	6.872		
15	1.226	4.358	6.762		
16	0.857	4.3044	7.351		
17	1.379	4.412	6.791		
18	0.561	3.980	7.141		
19	1.045	3.447	6.857		

This data creates an interesting graph. We plot the list of points and connect them together. This can be done using `ListLinePlot`.

```
ListLinePlot[xSample, Mesh -> All]
```

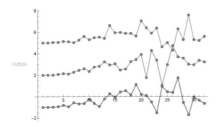

Note that since `xSample` is defined as a function depending on random values, each time we perform the line `ListLinePlot[xSample, Mesh -> All]` we get a different plot.

Finally, we can export the data and save them as an XLS file (among other formats). The built-in function `Export` will do the job.

```
?Export
```

Export["file.ext",expr] exports data to a file, converting it to the format corresponding to the file extension ext.

Export[file,expr,"format"] exports data in the specified format.

We save the data in the list `xSample` into a spreadsheet on the Desktop folder of the computer.

```
Export["/Users/hazrat/Desktop/xsample", xSample,
    "XLS"];
```

In fact, we can create a spreadsheet with several pages, as the following code shows.

```
Export["/Users/hazrat/Desktop/xsample", {"First Page" -> xSample,
    "Second Page" -> Transpose[xSample]}, "XLS"];
```

This will create a spreadsheet file named *xsample* in the Desktop folder. This spreadsheet has two pages, named First Page and Second Page. To demonstrate the function `Import`, we are going to open this file, get the data from the first page and then, using `Manipulate`, present the data dynamically.

The command `ss=Import["/Users/hazrat/Desktop/xSample"]` will import data of both pages of the spreadsheet into two lists and assign it to `ss`. If we would only like to have the first page of the spreadsheet (i.e., the first list), we can simply use `s[[1]]`. We will discuss how to access the elements of the lists in Chapter 4. The other option is to directly import the first page of the spreadsheet as follows

```
ss = Import["/Users/hazrat/Desktop/xSample", {"Data", 1}]
```

Now, using `Manipulate`, we present the data.

```
Manipulate[
ListLinePlot[ss[[All, 1 ;; i]], PlotRange->{{0, 32}, {-8, 8}},
Mesh -> All], {i, 0, 32, 1}]
```

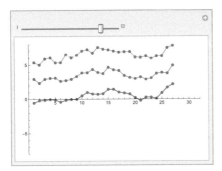

Mathematica provides built-in functions to access data on U.S. and other stocks, mutual funds and other financial instruments, as well as indices, and currency exchange rates. Once this data has been retrieved, one can use graphical capabilities of the software to present and analyse the data. Here we use `FinancialData` to retrieve the stock value of Facebook from November 2011 until the present and then plot it.

```
FinancialData["FB", "Nov. 20, 2011"]
{{{2012,5,18},38.23}, {{2012,5,21},34.03}, <<807>>,
  {{2015,8,7},94.3}}

DateListPlot[FinancialData["FB", "Nov. 20, 2011"]]
```

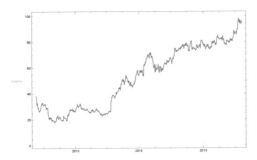

Next we plot the value of shares of Apple and Facebook in one graph.

```
DateListPlot[FinancialData[#, "Jan. 1, 2012"] & /@ {"FB", "AAPL"} ,
  PlotStyle -> {Blue, Gray}]
```

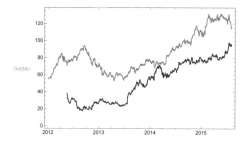

One can use graphical tools (appearing at the bottom of the graph) to decorate the output.

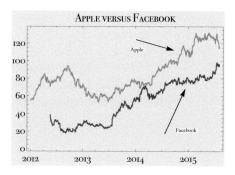

1.4 Linear algebra

The topic of linear algebra deals with systems of linear equations. We often
work with matrices (rectangles filled with numbers) in order to find solutions
of such systems. *Mathematica* can handle matrices and matrix computations
with ease. We will look at this topic in Chapter 13. Here we give a brief sample
of what we can do with *Mathematica*.

Consider the system of equation

$$\begin{cases} 3x + 8y + 9z = 8 \\ 7x - 8y - 9z = 44 \\ x + y - 3z = 0 \end{cases} \tag{1.1}$$

Represent the equation as

$$\begin{pmatrix} 3 & 8 & 9 \\ 7 & -8 & -9 \\ 1 & 1 & -3 \end{pmatrix} \begin{pmatrix} x \\ y \\ x \end{pmatrix} = \begin{pmatrix} 8 \\ 4 \\ 0 \end{pmatrix}.$$

We can enter a matrix into *Mathematica* by using Insert/Matrix from the
menu. Then we can calculate its determinant using the built-in function `Det`.

$$\mathbf{A} = \begin{pmatrix} 3 & 8 & 9 \\ 7 & -8 & -9 \\ 1 & 1 & -3 \end{pmatrix}$$

```
{{3, 8, 9}, {7, -8, -9}, {1, 1, -3}}
```

```
Det[A]
330
```

As can be seen from the above output, in *Mathematica* a matrix is represented by a list of lists (two layers of lists). Since the determinant of A is
non-zero, the matrix A is invertible and so it has an inverse matrix.

```
Inverse[A]
```

$$\left\{\left\{\frac{1}{10}, \frac{1}{10}, 0\right\}, \left\{\frac{2}{55}, -\frac{3}{55}, \frac{3}{11}\right\}, \left\{\frac{1}{22}, \frac{1}{66}, -\frac{8}{33}\right\}\right\}$$

```
Inverse[A] // MatrixForm
```

$$\begin{pmatrix} \frac{1}{10} & \frac{1}{10} & 0 \\ \frac{2}{55} & -\frac{3}{55} & \frac{3}{11} \\ \frac{1}{22} & \frac{1}{66} & -\frac{8}{33} \end{pmatrix}$$

Thus the solution to the system of equations 1.1 is $A^{-1}\begin{pmatrix} 8 \\ 4 \\ 0 \end{pmatrix}$. With the
help of *Mathematica* we get

```
Inverse[A].{8, 44, 0}
{26/5, -(116/55), 34/33}
```

Note that we have used the dot product . available in *Mathematica* to
multiply the matrices and vectors.

If we plot these equations which represent planes in space, they will intersect
at the point $\{26/5, -116/55, 34/33\}$.

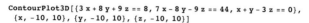

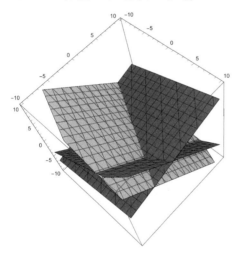

In fact, we could have used *Mathematica* to find the solution to this equation
by employing the built-in function `Solve`.

```
Solve[{3 x + 8 y + 9 z == 8, 7 x - 8 y - 9 z == 44,
    x + y - 3 z == 0}, {x, y, z}]

{{x -> 26/5, y -> -(116/55), z -> 34/33}}
```

By using a rule `->`, which we will study in Chapter 9, we will assign the
unique solution as the coordinates of a point. We plot a sphere at that point
and show this sphere and the three planes together using the command `Show`.

```
point = {x, y, z} /.
    {{x -> 26/5, y -> -(116/55), z -> 34/33}} // Flatten

{26/5, -(116/55), 34/33}
```

```
a = ContourPlot3D[{3 x + 8 y + 9 z == 8, 7 x - 8 y - 9 z == 44,
    x + y - 3 z == 0}, {x,-10,10}, {y,-10,10}, {z,-10,10}];
```

```
Show[a, Graphics3D[{Red, Sphere[point]}], Boxed -> False]
```

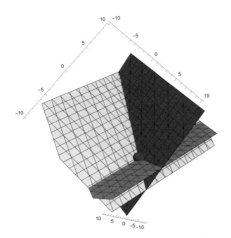

One can easily work with matrices symbolically. Here we show that for two 3×3 matrices A, B, $\mathrm{Tr}(AB) = \mathrm{Tr}(BA)$, where Tr is the trace of a matrix.

```
A = Array[x#1,#2 &, {3, 3}]
```
$\{\{x_{1,1},\ x_{1,2},\ x_{1,3}\},\ \{x_{2,1},\ x_{2,2},\ x_{2,3}\},\ \{x_{3,1},\ x_{3,2},\ x_{3,3}\}\}$

```
B = Array[y#1,#2 &, {3, 3}]
```
$\{\{y_{1,1},\ y_{1,2},\ y_{1,3}\},\ \{y_{2,1},\ y_{2,2},\ y_{2,3}\},\ \{y_{3,1},\ y_{3,2},\ y_{3,3}\}\}$

```
MatrixForm /@ {A, B}
```
$$\left\{ \begin{pmatrix} x_{1,1} & x_{1,2} & x_{1,3} \\ x_{2,1} & x_{2,2} & x_{2,3} \\ x_{3,1} & x_{3,2} & x_{3,3} \end{pmatrix}, \begin{pmatrix} y_{1,1} & y_{1,2} & y_{1,3} \\ y_{2,1} & y_{2,2} & y_{2,3} \\ y_{3,1} & y_{3,2} & y_{3,3} \end{pmatrix} \right\}$$

```
Tr[A.B]
```
$x_{1,1}\, y_{1,1} + x_{2,1}\, y_{1,2} + x_{3,1}\, y_{1,3} + x_{1,2}\, y_{2,1} + x_{2,2}\, y_{2,2} + x_{3,2}\, y_{2,3}$

```
Tr[A.B] == Tr[B.A]
```
True

1.5 Calculus

Roughly speaking, calculus is about studying the changes in a function. The two main ingredients are differentiation and integration and both are build upon the concept of the limit of a function approaching a certain point. *Mathematica* allows us to calculate limits, derivatives and integrals of functions. In fact, *Mathematica* can approach the calculation both numerically and symbolically. Here we give a quick overview of these capabilities. We study them in more detail in Chapter 15.

Let us start with the function $f(x) = x^2 \sin(1/x)$. This is a continuous function and we can ask *Mathematica* to calculate its limit as x approaches 0. The syntax in *Mathematica* to calculate the limit is close to how we present it in the mathematical literature,

$$\lim_{n \to 0} x^2 \sin(\frac{1}{x}).$$

```
Limit[x^2 Sin[1/x], x -> 0]
0
```

One can prove this by observing that $f(x)$ squeezes between $-x^2$ and x^2 and since these functions tend to 0 as x approaches to 0, the same will happen for $f(x)$. This is clear from the graph below.

```
Plot[{x^2, -x^2, x^2 Sin[1/x]}, {x, -0.05, 0.05}]
```

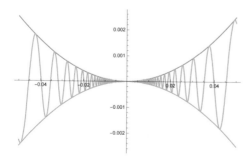

Not all functions are continuous at all points. Here is one example of such a function.

```
Limit[Sin[1/x], x -> 0]
Interval[{-1, 1}]
```

Next we look at differentiation. *Mathematica* can calculate the derivative of functions symbolically. We know for $y = \sin(x)$, we have $dy/dx = \cos(x)$. This is how we translate this into *Mathematica*.

```
D[Sin[x], x]
Cos[x]
```

Here is an impressive way to prove that this is actually correct, using the definition of differentiation.

```
FullSimplify[Limit[(Sin[x + h] - Sin[x])/h, h -> 0]]
Cos[x]
```

Another example where `FullSimplify` is needed to get the result in the simplest form is the following.

```
D[ArcTanh[Sin[x]], x]
Cos[x]/(1 - Sin[x]^2)

FullSimplify[D[ArcTanh[Sin[x]], x]]
Sec[x]
```

Next we are going to observe the behaviour of a function $f(x)$ and its derivative $f'(x)$, using the very useful `Manipulate` command. One can think of $f(x)$ as the speed of a vehicle and $f'(x)$ as its acceleration and how they change in the graph. The function we consider here is $f(x) = x^2 \sin(x)$.

We first calculate its derivative with respect to x, and then plot the function.

```
D[x^2 Sin[x], x]
x^2 Cos[x] + 2 x Sin[x]

Plot[x^2 Cos[x] + 2 x Sin[x], {x, -2 Pi, 2 Pi}]
```

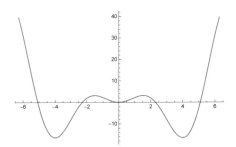

Now we put $f(x)$ and $f'(x)$ together using `Manipulate`.

```
Manipulate[
  Plot[{x^2 Sin[x], x^2 Cos[x] + 2 x Sin[x]}, {x, -2 Pi, k},
    PlotRange -> {{-2 Pi, 2 Pi}, {-30, 30}},
    PlotLegends -> "Expressions"], {k, -2 Pi + 1, 2 Pi}]
```

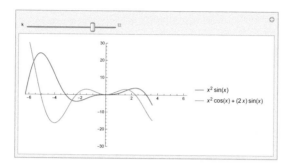

Next we demonstrate how to calculate the integral of a function using Integrate. We will calculate

$$\int_0^{\frac{\pi}{4}} \frac{x\cos(x)(x\sin(x) - \cos(x))}{1 + x\cos(x)} \, dx.$$

First, let us plot this function to get a better feel of it.

```
Plot[(x Cos[x] (x Sin[x] - Cos[x]))/(1 + x Cos[x]),
  {x, -2 Pi, 2 Pi}]
```

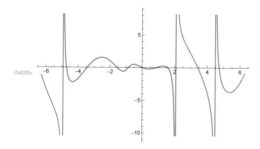

It is clear that in the region $-2\pi \le x \le 2\pi$, the denominator $1 + x\cos(x)$ will be zero for certain x and this region is presented as a vertical straight line in the plot. We find these points, i.e., we find those x such that $1 + x\cos(x) = 0$. Since this is not an algebraic equation, we use FindRoot to, well, find the roots of this equation.

```
Plot[1 + x Cos[x], {x, -2 Pi, 2 Pi}]
```

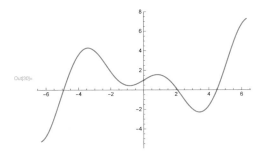

From the graph it is clear that the roots are close the points $\{-5, 2, 4.2\}$.

```
FindRoot[1 + x Cos[x] == 0, {x, #}] & /@ {-5, 2, 4.2}
{{x -> -4.91719}, {x -> 2.07393}, {x -> 4.48767}}
```

In fact, using Exclusions one can exclude an area from a graph, as the following shows:

```
Plot[(x Cos[x] (x Sin[x] - Cos[x]))/(1 + x Cos[x]),
  {x, -2 Pi, 2 Pi}, Exclusions -> {1 + x Cos[x ] == 0}]
```

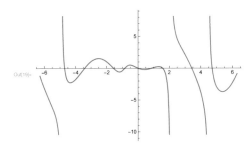

Before calculating the definite integral between the interval given, we ask *Mathematica* to calculate the indefinite integral. Notice how comfortable *Mathematica* is with symbolic computation. We make sure this calculation is correct. That is, if we calculate the derivative of the result we should get the original function. Here is the computation.

```
Integrate[(x Cos[x] (x Sin[x] - Cos[x])) / (1 + x Cos[x]), x]
```
$- x \, \text{Cos}[x] + \text{Log}[1 + x \, \text{Cos}[x]]$

```
D[- x Cos[x] + Log[1 + x Cos[x]], x]
```
$- \text{Cos}[x] + x \, \text{Sin}[x] + \dfrac{\text{Cos}[x] - x \, \text{Sin}[x]}{1 + x \, \text{Cos}[x]}$

```
Together[- Cos[x] + x Sin[x] + (Cos[x] - x Sin[x])/(1 + x Cos[x])]
```
$- \dfrac{x \, \text{Cos}[x] \, (\text{Cos}[x] - x \, \text{Sin}[x])}{1 + x \, \text{Cos}[x]}$

Finally, we calculate the definite integral both using the precise computation via Integrate and the numerical approach, using NIntegrate.

```
Integrate[(x Cos[x] (x Sin[x] - Cos[x])) / (1 + x Cos[x]), {x, 0, Pi/4}]
```
$- \dfrac{\pi}{4\sqrt{2}} + \text{Log}\left[\dfrac{1}{8}\left(8 + \sqrt{2}\,\pi\right)\right]$

```
NIntegrate[(x Cos[x] (x Sin[x] - Cos[x])) / (1 + x Cos[x]), {x, 0, Pi/4}]
```
$- 0.113653$

Finally, consider the function $2x^2 - x^3$. We would like to rotate this function around the y-axis and then calculate the volume generated below this generated surface.

```
Plot[2 x^2 - x^3, {x, 0, 2}]
```

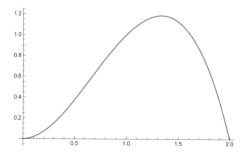

Now the command RevolutionPlot3D will rotate the graph.

```
RevolutionPlot3D[2 x^2 - x^3, {x, 0, 2}]
```

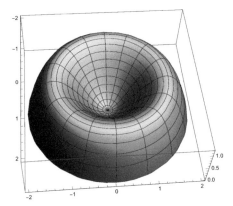

Finally, NIntegrate will calculate the desired volume for us.

```
Integrate[2 Pi x (2 x^2 - x^3), {x, 0, 2}]
(16 Pi])/5
```

2

Basics

This chapter introduces the basic capabilities of *Mathematica*, which include simple arithmetic, handling algebraic and trigonometric expressions and assigning values to variables. We will also look at dynamic objects, which allow us to see changes in the variables as they happen.

In this chapter we give a quick introduction to the very basic things one can do with Wolfram *Mathematica*®. We let the reader learn from reading the codes and avoid long and exhausting explanations, as the codes will speak for themselves.

2.1 *Mathematica* as a calculator

Mathematica can be used as a powerful calculator with the basic arithmetic operations; $+, -$ for addition and subtraction, $*, /$ for multiplication and division and $\char94$ for powers.

```
10^9 - 987654321
12345679

2682440^4 + 15365639^4 + 18796760^4
180630077292169281088848499041

20615673^4
180630077292169281088848499041
```

The last two calculations show that

$$2682440^4 + 15365639^4 + 18796760^4 = 20615673^4,$$

© Springer International Publishing Switzerland 2015
R. Hazrat, *Mathematica®: A Problem-Centered Approach*, Springer Undergraduate
Mathematics Series, DOI 10.1007/978-3-319-27585-7_2

disproving a conjecture of Euler that three fourth powers can never sum to a fourth power.[1]

Mathematica can handle large calculations:

```
2^9941-1
```

```
3460882824908512152429603957674133167226286689002385477904892834450062208098341144643643755441537075336644867476350501864147070933237397060837669040422926578964799370976035846955231904548491005030414980981854028350715968356223294196805976228133454473972084926090485519277062605491179359038906079598116383872143299427876363309537743819484486647112496768579888817221203300082146968446495614699719412692128433620646331385953757720046242420290646813260875582574884704893842439892702368849786430630930044229396033700105465953863020090730439444822025590974067005973305707995078329631309387398850801984162586351945229130425629366798595874957210311737477964188950607019417175060019371524300323636319342657985162360474512090898647074307803622983070381934454864937566479918042587755749738339033157350828910293923593527586171850199425548346718610745487724398807296062449119400666801128238240958164582617618617466040348020564668231437182554927847793809917495802552633233265364577438941508489539699028185300578708762293298033382857354192282590221696026655322108347896020516865460114667379813060562474800550717182503337375022673073441785129507385943306843408026982289639865627325971753720872956490728302897497713583308679515087108592167432185229188116706374484964985490944305412774440794079895398574694527721321665808857543604774088429133272929486968974961416149197398454328358943244736013876096437505146992150326837445270717186840918321709483693962800611845937461435890688111902531018735953191561073191960711505984880700270887058427496052030063194191166922106176157609367241948160625989032127984748081075324382632093913796444665700601391278360323002267409697776533329401822158003182889278072368602128982710306618115118964131893657845400296860012420391376964670183983594954112484565597312460737798777092071706710824503707457220155015899591766244957768006802482976673920392995410164224776445671222149803657927708412925555542817045572430846389988129960519227313987291200902060882060733762075892299473666405897427035811786879875694315078654420055603469625309399653955932310466430039146465805452965014040019423897552675534768248624631951431493188170905972588780111850281190559073677771187432814088678674286302108275149258477101296451833651979717375170900505673645964696355331369819296000267389583289299126738345726980325998955997501176664201042888546085699446442834195232948787488410595750197438786353119204210855804692460582533832967771946911459901921324984968810021189968284941331573164056304725480868921823442538195990383852412786840833479611419970101792978355653650755329138298654246225346827207503606740745956958127383748717825918527473164970582095181312905519242710280573023145554793628499010509296055849712377978984921839997037415897674154830708629145484724536724572622450131479992681684310464449439022250504859250834761894788889552527898400988196200014868575640233136509145628127191354858275083907891469979019426224883789463551
```

If a number of the form $2^n - 1$ happens to be prime, it is called a Mersenne prime. Recall that a *prime number* is a number greater than 1 which is divisible only by 1 and itself. It is easy to see that $2^2 - 1$, $2^3 - 1$ and $2^5 - 1$ are Mersenne primes. The list continues. In 1963, Gillies found that the above number, $2^{9941} - 1$, is a Mersenne prime. With my laptop it takes about 3 seconds for *Mathematica* to check that this is indeed a prime number.[2]

```
PrimeQ[2^9941-1]
True
```

Back to easier calculations:

```
24/17
24/17
```

Mathematica always tries to give a precise value, thus it returns back $\frac{24}{17}$ instead of attempting to evaluate the fraction.

---

[1] This conjecture remained open for almost 200 years, until Noam Elkies at Harvard came up with the above counterexample in 1988.

[2] The largest Mersenne prime found so far is $2^{57885161} - 1$ which was discovered in Jan 2013 and has $17,425,170$ digits.

```
Sin[Pi/5]
```

$$\frac{1}{2}\sqrt{\frac{1}{2}(5-\sqrt{5})}$$

Mathematica gives $\frac{1}{2}\sqrt{\frac{1}{2}(5-\sqrt{5})}$ as the value of $\sin(\pi/5)$, which is the precise value. This shows that *Mathematica* is not approaching the expressions numerically.

In order to get the numerical value, one can use the function N.

```
N[24/17]
1.41176
```

```
N[24/17, 20]
1.4117647058823529412
```

```
?N
N[expr] gives the numerical value of expr. N[expr, n] attempts to
give a result with n-digit precision.
```

As the above line shows, in order to get a quick description of a command, use ?Command. If you need further explanation use ??Command. If you need even more, use *Mathematica*'s Document Center in the Help Menu and search for the command. *Mathematica* comes with an excellent Help which contains explanations and many nice examples demonstrating how to use each of the functions available in this software.

All elementary mathematical functions are available here, Log, Exp, Sqrt, Sin, Cos, Tan, ArcSin,.... Here we evaluate

$$\sqrt{(\frac{\pi}{4})^2 + (0.5\,\text{Log}[2])^2},$$

```
Sqrt[(Pi/4)^2 + (0.5 Log[2])^2]
0.858466
```

We know that $\sin^2(x) + \cos^2(x) = 1$. We check this using *Mathematica*.

```
Sin[x]^2 + Cos[x]^2
Cos[x]^2 + Sin[x]^2
```

Apparently *Mathematica* has not recognised the identity. We try Simplify and TrigExpand to see if we can simplify the expression.

```
?Simplify
Simplify[expr] performs a sequence of algebraic and other
transformations on expr, and returns the simplest form it finds.
```

```
Simplify[Sin[x]^2 + Cos[x]^2]
1
```

```
?TrigExpand
TrigExpand[expr] expands out trigonometric functions in expr. >>

TrigExpand[Sin[x]^2 + Cos[x]^2]
1
```

This example introduces the commands `Simplify` and `FullSimplify`. If one is not happy with the result, one can always use `Simplify` and even `FullSimplify` to have *Mathematica* work a bit harder to come up with more simplification.

```
?FullSimplify
FullSimplify[expr] tries a wide range of transformations on expr
involving elementary and special functions, and returns the
simplest form it finds.
```

____ Problem 2.1

Show that

$$\frac{(1+\sqrt{5})^{10} - (1-\sqrt{5})^{10}}{1024\,\sqrt{5}} = 55.$$

$\Longrightarrow$ SOLUTION.

If we just enter the expression on the left-hand side of the equality into *Mathematica*, we will not get the result we are after (try it!). Thus we will use `Simplify` to get a more polished answer.

```
Simplify[((1 + Sqrt[5])^10 - (1 - Sqrt[5])^10)/(1024 Sqrt[5])]
55
```

The following problem shows that sometimes one needs to *investigate* a bit more in order to get the desired format of the expressions.

____ Problem 2.2

Use *Mathematica* to show that

$$\tan\frac{3\pi}{11} + 4\sin\frac{2\pi}{11} = \sqrt{11}.$$

$\Longrightarrow$ SOLUTION.

We enter the expression into *Mathematica*.

$$\text{Tan}\left[3\,\text{Pi}\,/\,11\right] + 4\,\text{Sin}\left[2\,\text{Pi}\,/\,11\right]$$

$$\text{Cot}\left[\frac{5\,\pi}{22}\right] + 4\,\text{Sin}\left[\frac{2\,\pi}{11}\right]$$

Clearly we didn't get $\sqrt{11}$ as an answer. We ask *Mathematica* to try a bit harder by using `Simplify`.

$$\text{Simplify}\left[\text{Tan}\left[3\,\text{Pi}\,/\,11\right] + 4\,\text{Sin}\left[2\,\text{Pi}\,/\,11\right]\right]$$

$$\text{Cot}\left[\frac{5\,\pi}{22}\right] + 4\,\text{Sin}\left[\frac{2\,\pi}{11}\right]$$

Since we didn't get what we were looking for, we try `TrigExpand`.

$$\text{TrigExpand}\left[\text{Tan}\left[3\,\text{Pi}\,/\,11\right] + 4\,\text{Sin}\left[2\,\text{Pi}\,/\,11\right]\right]$$

$$\frac{2\,\text{Cos}\left[\frac{\pi}{22}\right]}{5\,\text{Cos}\left[\frac{\pi}{22}\right]^4 \text{Sin}\left[\frac{\pi}{22}\right] - 10\,\text{Cos}\left[\frac{\pi}{22}\right]^2 \text{Sin}\left[\frac{\pi}{22}\right]^3 + \text{Sin}\left[\frac{\pi}{22}\right]^5} +$$

$$\frac{\text{Cos}\left[\frac{\pi}{22}\right]^5}{5\,\text{Cos}\left[\frac{\pi}{22}\right]^4 \text{Sin}\left[\frac{\pi}{22}\right] - 10\,\text{Cos}\left[\frac{\pi}{22}\right]^2 \text{Sin}\left[\frac{\pi}{22}\right]^3 + \text{Sin}\left[\frac{\pi}{22}\right]^5} -$$

$$\frac{2\,\text{Cos}\left[\frac{\pi}{22}\right]^9}{5\,\text{Cos}\left[\frac{\pi}{22}\right]^4 \text{Sin}\left[\frac{\pi}{22}\right] - 10\,\text{Cos}\left[\frac{\pi}{22}\right]^2 \text{Sin}\left[\frac{\pi}{22}\right]^3 + \text{Sin}\left[\frac{\pi}{22}\right]^5} -$$

$$\frac{10\,\text{Cos}\left[\frac{\pi}{22}\right]^3 \text{Sin}\left[\frac{\pi}{22}\right]^2}{5\,\text{Cos}\left[\frac{\pi}{22}\right]^4 \text{Sin}\left[\frac{\pi}{22}\right] - 10\,\text{Cos}\left[\frac{\pi}{22}\right]^2 \text{Sin}\left[\frac{\pi}{22}\right]^3 + \text{Sin}\left[\frac{\pi}{22}\right]^5} +$$

$$\frac{72\,\text{Cos}\left[\frac{\pi}{22}\right]^7 \text{Sin}\left[\frac{\pi}{22}\right]^2}{5\,\text{Cos}\left[\frac{\pi}{22}\right]^4 \text{Sin}\left[\frac{\pi}{22}\right] - 10\,\text{Cos}\left[\frac{\pi}{22}\right]^2 \text{Sin}\left[\frac{\pi}{22}\right]^3 + \text{Sin}\left[\frac{\pi}{22}\right]^5} +$$

$$\frac{5\,\text{Cos}\left[\frac{\pi}{22}\right] \text{Sin}\left[\frac{\pi}{22}\right]^4}{5\,\text{Cos}\left[\frac{\pi}{22}\right]^4 \text{Sin}\left[\frac{\pi}{22}\right] - 10\,\text{Cos}\left[\frac{\pi}{22}\right]^2 \text{Sin}\left[\frac{\pi}{22}\right]^3 + \text{Sin}\left[\frac{\pi}{22}\right]^5} -$$

$$\frac{252\,\text{Cos}\left[\frac{\pi}{22}\right]^5 \text{Sin}\left[\frac{\pi}{22}\right]^4}{5\,\text{Cos}\left[\frac{\pi}{22}\right]^4 \text{Sin}\left[\frac{\pi}{22}\right] - 10\,\text{Cos}\left[\frac{\pi}{22}\right]^2 \text{Sin}\left[\frac{\pi}{22}\right]^3 + \text{Sin}\left[\frac{\pi}{22}\right]^5} +$$

$$\frac{168\,\text{Cos}\left[\frac{\pi}{22}\right]^3 \text{Sin}\left[\frac{\pi}{22}\right]^6}{5\,\text{Cos}\left[\frac{\pi}{22}\right]^4 \text{Sin}\left[\frac{\pi}{22}\right] - 10\,\text{Cos}\left[\frac{\pi}{22}\right]^2 \text{Sin}\left[\frac{\pi}{22}\right]^3 + \text{Sin}\left[\frac{\pi}{22}\right]^5} -$$

$$\frac{18\,\text{Cos}\left[\frac{\pi}{22}\right] \text{Sin}\left[\frac{\pi}{22}\right]^8}{5\,\text{Cos}\left[\frac{\pi}{22}\right]^4 \text{Sin}\left[\frac{\pi}{22}\right] - 10\,\text{Cos}\left[\frac{\pi}{22}\right]^2 \text{Sin}\left[\frac{\pi}{22}\right]^3 + \text{Sin}\left[\frac{\pi}{22}\right]^5}$$

We try to simplify this expression yet again.

```
Simplify[TrigExpand[Tan[3 Pi / 11] + 4 Sin[2 Pi / 11]]]
```

$$\frac{16\left(\mathbb{i} + 2\,(-1)^{3/22} - 2\,(-1)^{7/22} + 2\,(-1)^{15/22} + 2\,(-1)^{17/22} + (-1)^{21/22}\right)}{\left(-1 + (-1)^{1/11}\right)^5 \left(1 - 10\,\mathrm{Cot}\!\left[\frac{\pi}{22}\right]^2 + 5\,\mathrm{Cot}\!\left[\frac{\pi}{22}\right]^4\right)}$$

Although it is an improvement, we have not yet got $\sqrt{11}$. We try again, this time with FullSimplify.

```
FullSimplify[TrigExpand[Tan[3 Pi / 11] + 4 Sin[2 Pi / 11]]]
```

$\sqrt{11}$

We thus obtained the result we were after. We could also have asked *Mathematica* directly whether the right-hand side of the identity is the same as the left-hand side.

```
Tan[3 Pi / 11] + 4 Sin[2 Pi / 11] == Sqrt[11]
```

$\mathrm{Cot}\!\left[\dfrac{5\,\pi}{22}\right] + 4\,\mathrm{Sin}\!\left[\dfrac{2\,\pi}{11}\right] == \sqrt{11}$

```
FullSimplify[Tan[3 Pi / 11] + 4 Sin[2 Pi / 11] == Sqrt[11]]
```
True

∟

For a complete list of elementary functions have a look at Functional Navigator: Mathematics and Algorithms: Mathematical Functions in *Mathematica*'s Help.

Exercise 2.1
Show that $\sqrt{\sqrt[3]{64}\left(2^2 + (1/2)^2\right) - 1} = 4$.

Exercise 2.2
Show that

$$\left(\frac{1}{2} + \cos\left(\frac{\pi}{20}\right)\right)\left(\frac{1}{2} + \cos\left(\frac{3\pi}{20}\right)\right)$$
$$\left(\frac{1}{2} + \cos\left(\frac{9\pi}{20}\right)\right)\left(\frac{1}{2} + \cos\left(\frac{27\pi}{20}\right)\right) = \frac{1}{16}.$$

───── Problem 2.3

Using *Mathematica*, explain why $4 + 6/4 * 3\hat{\ } - 2 + 1 = \dfrac{31}{6}$.

⟹ SOLUTION.

If you look at *Mathematica*'s Help, under "Special Ways to Input Expressions", you will see the following note: "The *Mathematica* language has a definite grammar which specifies how your input should be converted to internal form." One aspect of the grammar is that it specifies how pieces of your input should be grouped. The general rule is that if $\otimes$ has higher precedence than $\oplus$, then $a \oplus b \otimes c$ is interpreted as $a \oplus (b \otimes c)$, and $a \otimes b \oplus c$ is interpreted as $(a \otimes b) \oplus c$. You will then find a long table listing which operation has a higher precedence and thus, based on that, you will be able to explain why $4 + 6/4 * 3\hat{\ } - 2 + 1$ amounts to $\frac{31}{6}$.

However, common sense tells us that instead of creating an ambiguous expression such as $4 + 6/4 * 3\hat{\ } - 2 + 1$, one should use parentheses () to group objects together and make the expression more clear. For example, one could write $4 + \left((6/4) * 3\hat{\ }(-2) \right) + 1$, or even better use *Mathematica*'s Palette (Basic Math Assistance) and type

$$4 + \frac{6}{4} \times 3^{-2} + 1.$$

└─────────────

♣ TIPS

– The mathematical constant e, the exponential number, is defined in *Mathematica* as E, or Exp. To get e^n use either E^ n or Exp[n]. The constant π can be typed as Pi.

– Comments can be added to the codes using (* comment *).

```
(* the most beautiful theorem in Mathematics *)
E^(I Pi) + 1
0
```

– The symbol % refers to the previous output produced. %% refers to the second previous output, and so on.

– If in calculations you don't get what you are expecting, use **Simplify** or even **FullSimplify** (see Problem 2.2).

– To get a numerical approximation, use N[expr] or alternatively, expr//N (see Problem 3.3 for different ways of applying a function to a variable).

Use `EngineeringForm[expr,n]` and `ScientificForm[expr,n]` to get other forms of numerical approximations to n significant digits.

2.2 Numbers

There are several standard ways to start with an integer and produce new numbers out of it. For example, starting from 4, one can form $4 \times 3 \times 2 \times 1$, which is represented by 4!.

```
4!
24
```

```
123!
12146304367025329675766243241881295855454217088483383231532891
81618292358923621676688311569606126402021707358352212940477782
591091570411651472186029519906261646730733907419814952960000
00000000000000000000000000000
```

The fundamental theorem of arithmetic states that one can decompose any natural number n as a product of powers of primes and this decomposition is unique, i.e., $n = p_1^{k_1} \cdots p_t^{k_t}$ where p_i's are prime. Thus $12 = 2^2 \times 3^1$ and $37534 = 2 \times 7^2 \times 383$. *Mathematica* can do all of these:

```
FactorInteger[12]
{{2, 2}, {3, 1}}
```

```
FactorInteger[37534]
{{2,1},{7,2},{383,1}}
```

```
2^1 * 7^2 * 383^1
37534
```

```
FactorInteger[6473434456376432]
{{2,4},{3239053,1},{124909859,1}}
```

```
PrimeQ[124909859]
True
```

```
Prime[8]
19
```

`Prime[n]` produces the n-th prime number. `PrimeQ[n]` determines whether n is a prime number. More than 2200 years ago Euclid proved that the set of prime numbers is infinite. His proof is used even today in modern books. However, it is not that long ago that we also learned that there is no simple formula that produces only prime numbers.

In 1640 Fermat conjectured that the formula $2^{2^n}+1$ always produces a prime number. Almost a hundred years later the first counterexample was found.

```
PrimeQ[2^(2^1)+1]
True

PrimeQ[2^(2^2)+1]
True

PrimeQ[2^(2^3)+1]
True

PrimeQ[2^(2^4)+1]
True

PrimeQ[2^(2^5)+1]
False

2^(2^5)+1
4294967297

FactorInteger[2^(2^5)+1]
{{641,1},{6700417,1}}
```

This shows that $2^{2^5} + 1$ is not a prime number. In fact it decomposes into two prime numbers $2^{2^5} + 1 = 641 \times 7600417$.

▬▬▬ Problem 2.4

What is the probability that a randomly chosen 13-digit number will be a prime?

$\Longrightarrow$ SOLUTION.

The probability is the number of 12-digit prime numbers over the number of all 12-digit numbers. So we start by finding how many 12-digit numbers exist:

```
10^13 - 10^12
9000000000000
```

Next, we will find how many 13-digit prime numbers exist. We will use the following built-in function of *Mathematica*.

```
?PrimePi

PrimePi[x] gives the number of primes less than
or equal to x. >>

PrimePi[10^13]
346065536839

PrimePi[10^12]
37607912018

N[(346065536839 - 37607912018)/9000000000000]*100
3.42731
```

So the probability that we randomly pick a 12-digit prime number is only 3.42 percent.

▌▬▬▬▬▬▬▬▬

Exercise 2.3

Check that 123456789098765432111 is a prime number.

Exercise 2.4

Check that for any given positive integer $n \geq 3$, the least common multiple of the numbers $1, 2, \cdots, n$ is greater than 2^{n-1}. (Hint: see LCM.)

Exercise 2.5

Check that the number 32! ends with 7 zeros.

▬▬▬▬ ## Problem 2.5

The function Mod[m,n] gives the remainder on division of m by n. Let $a = 98$ and $b = 75$. Show that $a^a b^b$ ends with exactly 98 zeros.

⟹ SOLUTION.

Clearly you can directly compute $a^a b^b$ and count the number of zeros as in Exercise 2.5. But this time you have to count whether the number ends with 98 zeros. Here is a more clever approach: Check whether the remainder of this number when divided by 10^{98} is zero, but its remainder on division by 10^{99} is not.

```
Mod[98^98 * 75^75, 10^98]
0

Mod[98^98 * 75^75, 10^99]
5000000000000000000000000000000000000000000000000000000000
00000000000000000000000000000000000000000
```

▌▬▬▬▬▬▬▬▬

▬▬▬▬ ## Problem 2.6

Recall that for integers m and n, the binomial coefficient $\begin{pmatrix} n \\ m \end{pmatrix}$ is defined as $\dfrac{n!}{m!(n-m)!}$. Using *Mathematica*, check that

$$\begin{pmatrix} m+n \\ m \end{pmatrix} = \frac{(m+n)!}{m!n!}.$$

$\Longrightarrow$ SOLUTION.

Not only is the binomial function available in *Mathematica*, but *Mathematica* can also perfectly handle it symbolically, as the solution to this problem shows. We will talk more about symbolic computations in Section 2.3.

```
Binomial[m + n, n] == (n + m)!/(n! m!)
Binomial[m + n, n] == (m + n)!/(m! n!)

FullSimplify[Binomial[m + n, n] == (n + m)!/(n! m!)]
True
```

This is another instance where we need to use `FullSimplify` to make *Mathematica* work harder to come up with the result.

We will discuss the different equalities available in *Mathematica* in Section 2.6. However, for the time being, note that `==` is used to compare both sides of equations.

There are several more integer functions available in *Mathematica*, which can be found in Functional Navigator: Mathematics and Algorithms: Mathematical Functions: Integer Functions.

♣ TIPS

- The command `NextPrime[n]` gives the next prime larger than `n` and `PrimePi[n]` gives the number of primes less than or equal to `n` (see Problem 2.4).

- For integers m and n, one can find unique numbers q and r such that r is positive, $m = qn + r$ and $r < |q|$. Then `Mod[m,n]=r` and `Quotient[m,n]=q`.

- If an evaluation is taking a long time, in order to stop the evaluation use Alt+. (for Windows) and Cmd+. (for Apple Macintosh). For example, try to calculate the 1234567891011-th prime number. If you can't wait to get the result, you now know how to stop the process. There are cases where pressing Alt+. does not help, even if you do it several times. In these situations, use the Evaluation menu and choose Quit Kernel.

2.3 Algebraic computations

One of the abilities of *Mathematica* is to handle symbolic computations, i.e., *Mathematica* can comfortably work with symbols (we have seen one example of this in Problem 2.6). Consider the expression $(x + 1)^2$. One can use *Mathematica* to expand this expression:

```
Expand[(1 + x)^2]
1 + 2 x + x^2
```

Mathematica can also do the inverse of this task, namely factorise an expression:

```
Factor[1 + 2 x + x^2]
(1 + x)^2
```

While expansion of an algebraic expression is a simple and routine procedure, the factorization of algebraic expressions is often quite challenging. My favorite example is this one. Try to factorise the expression $x^{10} + x^5 + 1$. Here is one way to do it:

$$x^{10} + x^5 + 1 \quad \text{(adding } x^i - x^i, 1 \leq i \leq 9, \text{ to the expression we have)}$$
$$= x^{10} + \underbrace{x^9 - x^9}_{} + \underbrace{x^8 - x^8}_{} + \cdots + \underbrace{x^6 - x^6}_{} +$$
$$+ \underbrace{x^5 - x^5}_{} + x^5 + \underbrace{x^4 - x^4}_{} + \cdots + \underbrace{x - x}_{} + 1 \quad \text{(now rearranging the terms)}$$
$$= x^{10} + x^9 + x^8 - x^9 - x^8 - x^7 + x^7 + x^6 + x^5 - x^6 - x^5 - x^4$$
$$+ x^5 + x^4 + x^3 - x^3 - x^2 - x + x^2 + x + 1$$
$$= x^8(x^2 + x + 1) - x^7(x^2 + x + 1) + x^5(x^2 + x + 1) - x^4(x^2 + x + 1)$$
$$+ x^3(x^2 + x + 1) - x(x^2 + x + 1) + x^2 + x + 1$$
$$= (x^2 + x + 1)(x^8 - x^7 + x^5 - x^4 + x^3 - x + 1)$$

Mathematica can easily come up with this factorization:

```
Factor[x^10 + x^5 + 1]
(1 + x + x^2) (1 - x + x^3 - x^4 + x^5 - x^7 + x^8)
```

━━━━ Problem 2.7

Prove that the product of four consecutive numbers plus one is always a square number.

⟹ SOLUTION.

We first check this is indeed the case for an example:

```
Sqrt[13*14*15*16 + 1]
209
```

Now here is a proof:

```
Factor[n (n + 1) (n + 2) (n + 3) + 1]
(1 + 3 n + n^2)^2
```

L_____

_____ Problem 2.8

Here is a formula to generate many instances where the sum of three fourth powers is a fourth power. Check this with *Mathematica* and find some examples of such numbers.

$$(85v^2 + 484v - 313)^4 + (68v^2 - 586v + 10)^4 + (2u)^4 = (357v^2 - 204v + 363)^4,$$

where
$$u^2 = 22030 + 28849v - 56158v^2 + 36941v^3 - 31790v^4.$$

$\Longrightarrow$ SOLUTION.

This solution is not optimal. We give another solution to this in Exercise 10.2 in Chapter 10 based on the pattern matching approach.

We use **Sqrt** to write u in terms of v and then enter the formula into *Mathematica*, asking it to compare the two sides of the equality using ==.

```
u = Sqrt[22030 + 28849 v - 56158 v^2 +  36941 v^3 - 31790 v^4];

(85 v^2 + 484 v - 313)^4 + (68 v^2 - 586 v + 10)^4 + (2 u)^4 ==
(357 v^2 - 204 v + 363)^4

(10 - 586 v + 68 v^2)^4 + (-313 + 484 v + 85 v^2)^4 +
16 (22030 + 28849 v - 56158 v^2 + 36941 v^3 -
     31790 v^4)^2 == (363 - 204 v + 357 v^2)^4
```

Since we didn't get an answer, we use **Simplify** to make *Mathematica* work harder.

```
Simplify[(85 v^2 + 484 v - 313)^4 +
(68 v^2 - 586 v + 10)^4 + (2 u)^4 == (357 v^2 - 204 v + 363)^4]
True
```

L_____

♣ TIPS

– The command **Together** converts a sum of terms into a single term over a common denominator. The command **Apart** (almost) does the reverse of **Together** (see Exercise 2.7).

Exercise 2.6

Factorise the polynomial $(1 + x)^{30} + (1 - x)^{30}$.

Exercise 2.7

Using `Together`, write the expression

$$\frac{1}{1 + x} + \frac{1}{1 + \frac{1}{1+x}}$$

with a single denominator. Now apply `Apart` to the result to get an expression as a sum of terms with minimal denominators.

There are several more algebraic functions available in *Mathematica*, which can be found in Functional Navigator: Mathematics and Algorithms: Mathematical Functions: Polynomial Algebra.

2.4 Trigonometric computations

Similar to algebraic expressions (Section 2.3), *Mathematica* can handle trigonometric expressions, as we saw in Problem 2.2. Here one uses `TrigExpand` and `TrigFactor` to work with trig. expressions.

Mathematica is quite at ease with trig. identities, as the following problem demonstrates.

▬▬ Problem 2.9

Using *Mathematica*, check that the following trigonometric identities hold:

$$\sin^3(x)\cos^3(x) = \frac{3\sin(2x) - \sin(6x)}{32}$$

$$\frac{1 + \sin(x) - \cos(x)}{1 + \sin(x) + \cos(x)} = \tan(x/2)$$

⟹ SOLUTION.

The only challenge here is to translate these expressions correctly into *Mathematica*.

```
Simplify[Sin[x]^3 Cos[x]^3 == (3 Sin[2 x] - Sin[6 x])/32]
True
```

```
Simplify[(1 + Sin[x] - Cos[x])/(1 + Sin[x] + Cos[x]) == Tan[x/2]]
True
```

Note that == is used to compare both sides of equations. We will discuss the different equalities available in *Mathematica* in Section 2.6.

∟▬▬▬▬▬▬▬▬▬

Exercise 2.8

Using *Mathematica*, show that

$$\frac{1 + \sin(x) - \cos(x)}{1 + \sin(x) + \cos(x)} = \tan(x/2).$$

♣ TIPS

– The argument of trig. functions, e.g., Sin, is assumed to be in radians. (Multiply by Degree to convert from degrees to radians.)

```
Sin[30 Degree]
1/2
```

2.5 Variables

In order to feed data into a computer program one needs to define variables to be able to assign data to them. As long as you use common sense, any names you choose for variables are valid in *Mathematica*. Names like x, y, x3, myfunc, xQuaternion,... are all fine. Do not use an underscore _ to define a variable.[3] Also note that *Mathematica* is case sensitive, thus xy and xY are considered as two different variables.

```
x = 3
3

y = 4
4

x^2 + y^2
25

Sqrt[x^2 + y^2]
5
```

If we need to enter several statements in one line, we can separate them with ;.

---

[3] This is quite common in Pascal or C, to define variables such as x_printer, com_graph,.... In *Mathematica*, the underscore is reserved and will be used in the definition of functions in Chapter 3.

```
t = 7; s = 4; t!/(s! (t - s)!)
35
```

One can assign any type of data to a variable. Not only would *Mathematica* not complain, but she would also carry out all the computations with that specific data. Here we drag a picture to the *Mathematica* front end (editor) and assign it to the variable x.

x =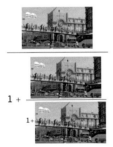

```
x/(1 + x/(1 + x))
```

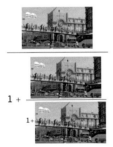

━━━ Problem 2.10

Using Expand, for Expand[(1+x)^ 2], instead of obtaining $1 + 2x + x^2$ we get

```
Expand[(1 + x)^2]
16
```

What seems to be the problem?

$\Longrightarrow$ SOLUTION.

If you are working through this section, at the beginning of this section you will already have defined x=3. Thus *Mathematica* will take this into account when working with the expression $(1 + x)^2$, which then amounts to 16. This demonstrates one of the common mistakes one tends to make in *Mathematica*, namely using variables which have already been defined, as undefined symbols. In order to clear the value or definition of a variable, use Clear.

```
Clear[x]

Expand[(1 + x)^2]
1 + 2 x + x^2
```

♣ TIPS

– Use Clear[x] to clear the value given to the variable x, before using x as a symbol.

– Use Clear["Global`*"] to clear values and definitions given to *all the* symbols.

– Assigning a value to a symbol works globally. That means, if you open a new NoteBook, the values given to variables in a previous NoteBook still exist.

2.6 Equalities =, :=, ==

Primarily there are three equalities in *Mathematica*, =, := and ==. There is a fundamental difference between = and :=. Study the following example:

```
x=5;y=x+2;

y
7

x=10
10
y
7

x=15
15

y
7
```

So changing the value of x does not affect the value of y. Now compare this with the following example, where we replace = with := in the definition of y.

```
x=5;y:=x+2;
```

```
y
7
```

```
x=10
10
```

```
y
12
```

```
x=15
15
```

```
y
17
```

From the first example it is clear that when we define y=x+2 then y takes the *value* of x+2 and this will be assigned to y. No matter if x changes its value, the value of y remains the same. In other words, y is independent of x. But in y:=x+2, y is dependent on x, and when x changes, the value of y changes too. That is using := makes y a function with variable x. The following is an excellent example demonstrating the difference between = and :=.

```
?Random
```

```
Random[ ] gives a uniformly distributed pseudorandom Real in the
range 0 to 1.
```

```
x=Random[]
0.246748
```

```
x
0.246748
```

```
x
0.246748
```

```
x:=Random[]
```

```
x
0.60373
```

```
x
0.289076
```

```
x
0.564378
```

When defining x=Random[], the function Random generates a number and this number will be assigned to x. Each time we call on x, this number is what we get. However, when we define x:=Random[], then the definition of x is

`Random[]`. Thus when we call x, we have in fact called on `Random` which then generates a new random number.

We will examine this difference between = and := again in Example 4.7. Finally, the equality == is used to compare:

```
5==5
True

3==5
False
```

We will discuss this further in Section 6.1.

2.7 Dynamic variables

The new version of *Mathematica*[4] comes with an ability to define *dynamic* variables. This means one can monitor the changes in a variable "live", i.e., as they happen. We are going to introduce this feature early in the book to take advantage of it as we go along.

We saw in Section 2.5 that one can define variables and assign values to them.

```
x = 3
3

x = 10
10
```

Here when we assign 10 to x, although this is the new value of x, in the line above it, i.e., x=3, 3 does not change. However, if we define the variable x as a dynamic variable, then each time we change the value of x anywhere in the program, all the old values also change to the new value accordingly.

```
Dynamic[x]
10
```

Then if in the next line we change the value to x=15, we will see that the value of the previous line immediately changes to 15 as well.

```
Dynamic[x]
15

x=15
15
```

One can control the value of the variable x by introducing a *slider*.

---
[4] Currently version 10.

`Slider[Dynamic[x]]`

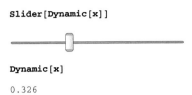

`Dynamic[x]`

0.326

You will see that as you drag the slider, the value of x changes. This already gives us a lot of power, as the following example will show. Recall from Section 2.3 that we can expand expressions using Expand.

```
Expand[(1 + y)^2]
1 + 2 y + y^2
```

```
Expand[(1 + y)^3]
1 + 3 y + 3 y^2 + y^3
```

Now we can simply consider $(1 + y)^n$ and then, defining n as a dynamic variable and controlling it with a slider, we can change the value of n by dragging the slider and see the expansions of $(1 + y)^n$ for different values of n as they happen right in front of our eyes!

`Slider[Dynamic[n], {1, 10, 1}]`

$1 + 7\,y + 21\,y^2 + 35\,y^3 + 35\,y^4 + 21\,y^5 + 7\,y^6 + y^7$

`Dynamic[Expand[(1 + y) ^n]]`

$1 + 7\,y + 21\,y^2 + 35\,y^3 + 35\,y^4 + 21\,y^5 + 7\,y^6 + y^7$

Note that, when defining Slider, the value of x varies from 0 to 1. If we want to change this interval, as in the previous example, we can specify the interval and the step that is added to x each time by using {xmin,xmax, step}.

A similar concept to Slider is the function Manipulate which allows us to change the value of a variable and see the result "live".

`Manipulate[Expand[(1 + y) ^n], {n, 1, 10, 1}]`

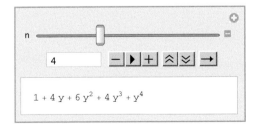

The control buttons can be used to start or stop the process, and make it faster or slower. Just try it out.

We will see later, for example in Chapter 14 when we deal with graphics, that we can use `Manipulate` to change the value of our parameters and see how the graph changes accordingly.

▬▬ Problem 2.11

Using `Manipulate`, observe that the polynomial $x^{2n}+x^n+1$ can be decomposed into smaller factors for any $1 \le n \le 20$ except $n = 1, 3, 9$.

⟹ SOLUTION.

```
Manipulate[Factor[x^(2 n) + x^n + 1], {n, 1, 20, 1}]
```

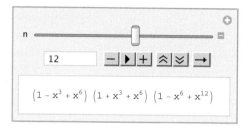

▬▬ Problem 2.12

Using `Manipulate`, find out for which positive integers n and m, between 1 and 100, $m^2 + n^2$ is a square number (these are called Pythagorean pairs).

⟹ SOLUTION.

Later in Chapter 8, when we are discussing loops, we will write a program to generate these numbers (Problem 8.12). Here we will use `Manipulate`, defining two dynamic variables m and n, and we will look at the result of $\sqrt{m^2 + n^2}$. If this is an integer, then (m, n) is a Pythagorean pair.

```
Manipulate[Sqrt[m^2 + n^2], {m, 1, 100, 1}, {n, 1, 100, 1}]
```

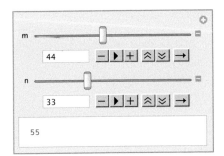

3
Defining functions

This chapter shows how to define functions in *Mathematica*. Examples of functions with several variables and anonymous functions are given.

Functions in mathematics define rules about how to handle data. A function is a rule which assigns to each element in its domain a unique element in a specific range. For example, the function f defined as $f(n) = n^2 + 1$ will receive as an input (a number) n and its output will be $n^2 + 1$. Besides the wide use of $f(n)$, there are several other ways to indicate how the rule f applies to n, such as $n \xrightarrow{f} n^2 + 1$, $nf = n^2 + 1$ or even $n^f = n^2 + 1$. We will see that Wolfram *Mathematica®*, besides supporting `f[n]`, has two other ways to apply a function to data, namely `f@n` and `n//f`.

3.1 Formulas as functions

Defining functions is one of the strong features of *Mathematica*. One can define a function in several different ways in *Mathematica* as we will see in the course of this chapter.

Let us start with a simple example, defining the formula $f(n) = n^2 + 4$ as a function and calculating $f(-2)$:

```
f[n_]:= n^2 +4
```

First notice that in defining a function we use :=. The symbol `n` is a dummy variable and, as expected, one plugs in the data in place of `n`.

© Springer International Publishing Switzerland 2015
R. Hazrat, *Mathematica®: A Problem-Centered Approach*, Springer Undergraduate
Mathematics Series, DOI 10.1007/978-3-319-27585-7_3

```
f[3]
10
```

```
f[-2]
8
```

In fact, as we will see later, one can plug "anything" in place of n and that is why functions in *Mathematica* are far superior to those in C and other languages.

```
f[3.56]
16.6736
```

```
f[elephant]
4 + elephant^2
```

One more note about the extra underscore _ in the definition of the function. The underscore, which will be called blank here, stands (or rather sits) for the expression which will be passed to f. So we cheated a little when we said we plug data in place of n. The underscore, named n, gets the data and f applies its rule to this data. This data can have any pattern. If this is confusing, forget this technicality now. We will talk about patterns and pattern matching in Chapter 10 and leave it as it is for the moment.

Here is yet another example of how one can pass practically anything into a function.

```
f[x_] := x/(1 + x)
f[f[x]]
```

$$\frac{x}{(1+x)\left(1+\frac{x}{1+x}\right)}$$

```
f[x_] := x/(1 + x)
f[f[f[x]]]
```

$$\frac{x}{(1+x)\left(1+\frac{x}{1+x}\right)\left(1+\frac{x}{(1+x)\left(1+\frac{x}{1+x}\right)}\right)}$$

```
Simplify[%]
```

$$\frac{x}{(1+3x)}$$

Now here is something even more amusing.

```
?CurrentImage
```

```
CurrentImage[] returns the current image captured from a
connected camera.
```

```
f[f[f[CurrentImage[]]]]
```

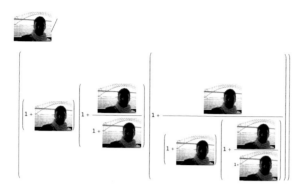

```
Simplify[%]
```

We proceed by defining the function $g(x) = x + \sin(x)$.

```
g[x_]:= x+Sin[x]
g[Pi]
```

π

It is very easy to compose functions in *Mathematica*, i.e., apply functions one after the other on data. Here is an example of this:

```
f[x_]:=x^2+1
g[x_]:=Sin[x]+Cos[x]

f[f[x]]
1 + (1 + x^2)^2

f[g[x]]
1+(Cos[x]+Sin[x])^2

g[f[x]]
Cos[1+x^2]+Sin[1+x^2]
```

This example clearly shows that the composition of functions is not a commutative operation, that is $fg \neq gf$.

_____ Problem 3.1

Using *Mathematica*, show that

$$\frac{1}{1+\frac{1}{1+\frac{1}{1+\frac{1}{1+x}}}} = \frac{3+2x}{5+3x}. \tag{3.1}$$

⟹ SOLUTION.

If we define $f(x) = \frac{1}{1+x}$, then $f(f(x)) = \frac{1}{1+\frac{1}{1+x}}$ and $f(f(f(x))) = \frac{1}{1+\frac{1}{1+\frac{1}{1+x}}}$.

This shows a way to capture the left-hand side of Equality 3.1 without going through the pain of typing it.

```
f[x_] := 1/(1 + x)

f[f[f[f[x]]]]
1/(1 + 1/(1 + 1/(1 + 1/(1 + x))))

Simplify[f[f[f[f[x]]]]]
(3 + 2 x)/(5 + 3 x)
```

In Section 8.3 we will come back to this example to show how one can systematically apply a function to itself multiple times.

_____ Problem 3.2

Define the function $f(x) = ||x|-1|$ in *Mathematica* and plot $f(x), f(f(x))$ and $f(f(f(x)))$. The absolute value function $|\ \ |$ is defined as Abs in *Mathematica*.

⟹ SOLUTION.

```
f[x_]:=Abs[Abs[x]-1]
```

Let us understand this function. Depending on the value of x, the absolute value function applies and makes the values positive. We get

$$f(x) = \begin{cases} x - 1 & \text{if } x \geq 1 \\ -x + 1 & \text{if } 0 \leq x < 1 \\ x + 1 & \text{if } -1 \leq x < 0 \\ -x - 1 & \text{if } x \leq -1 \end{cases}$$

The reader might imagine that for $f(f(x))$ one should consider several other cases. Once we have defined $f(x)$, we can ask *Mathematica* to plot them for us.

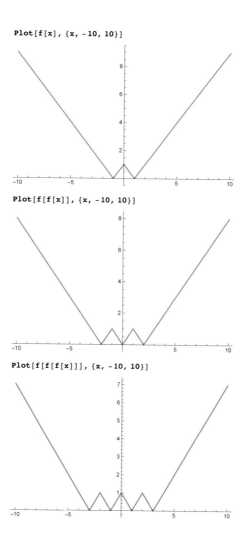

Plot[f[x], {x, -10, 10}]

Plot[f[f[x]], {x, -10, 10}]

Plot[f[f[f[x]]], {x, -10, 10}]

Exercise 3.1

Define $f(x) = \sqrt{1 + x}$ in *Mathematica* and show that

$$f(f(f(f(f(x))))) = \sqrt{1 + \sqrt{1 + \sqrt{1 + \sqrt{1 + \sqrt{1 + x}}}}}.$$

Recall that the Fibonacci sequence starts with the two terms, 1, 1 and the next term in the sequence is the sum of the previous two. Thus the first 7 Fibonacci numbers are $1, 1, 2, 3, 5, 8, 13$. The function `Fibonacci[n]` gives the n-th Fibonacci number.

```
Fibonacci[7]
13
```

Here is a little function which determines if the n-th Fibonacci number is divisible by 5 (see Problem 2.5 for the use of function `Mod`).

```
rf[n_]:=Mod[Fibonacci[n],5]
rf[14]
2

rf[15]
0
```

Thus the 15th Fibonacci number is divisible by 5. Note that the function `remain` is itself a composition of two functions, namely the functions `Fibonacci` and `Mod`.

Besides the traditional method of using `rf[x]`, there are two other ways to apply a function to an argument as follows:

```
15//rf
0

rf@15
0
```

──── Problem 3.3

Design a function to check whether for a number n, the formula $n!+1$ generates a prime number.

⟹ SOLUTION.

The function is a composition of two functions, first to generate `n!+1` and then using `PrimeQ` to test whether this number is prime.

```
pTest[n_] := PrimeQ[n! + 1]

pTest[2]
True

pTest[3]
True

pTest[4]
False
```

Here are the other ways to apply the function to a variable.

```
4//pTest
False

pTest@4
False
```

⌊▬▬▬▬▬▬▬▬▬

▬▬▬ Problem 3.4

Define the function

$$b(n) = 1 + \binom{n}{1} + \binom{n}{2} + \binom{n}{3},$$

and find out whether 2^{2008} is divisible by $b(23)$.

⟹ SOLUTION.

Recall that $\binom{n}{m}$ is defined by Binomial[n,m] in *Mathematica*.

```
b[n_] := Binomial[n, 1] + Binomial[n, 2] + Binomial[n, 3] + 1

b[23]
2048

Mod[2^2008, b[23]]
0
```

Recall from Problem 2.5 that Mod[m,n] returns the remainder on division of m by n. If this remainder is zero, it clearly means that m is divisible by n. There is also the command Divisible which takes care of this situation.

```
?Divisible

Divisible[n,m] yields True if n is divisible by m, and yields
False if it is not. >>

Divisible[2^2008, b[23]]
True
```

Later, in Problem 4.6, we will determine all the positive integers n between 3 and 50 for which 2^{2008} is divisible by $b(n)$ and will see that in fact there are very few n with this property.

⌊▬▬▬▬▬▬▬▬▬

In a similar manner, one can define functions of several variables. Here is a simple example defining $f(x, y) = \sqrt{x^2 + y^2}$.

```
f[x_,y_]:=Sqrt[x^2+y^2]
f[3,4]
5
```

───── Problem 3.5

Define the function $f(x,y) = ||x|-|y||$ in *Mathematica* and plot its graph for $-10 \le x, y \le 10$.

⟹ SOLUTION.

```
f[x_, y_] := Abs[Abs[x] - Abs[y]]

Plot3D[f[x, y], {x, -10, 10}, {y, -10, 10}]
```

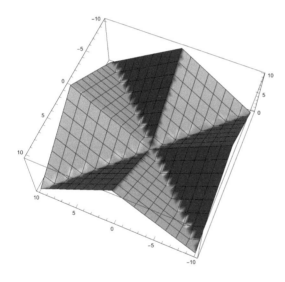

└────────────

───── Problem 3.6

Recall that if one wants to prove, by mathematical induction, that a statement $P(n)$ is valid for all natural numbers n, one needs first to check $P(1)$ is valid, and then assuming $P(k)$ is correct, prove that $P(k+1)$ is also valid.

Using *Mathematica* and mathematical induction, prove the following identities:

$$1 + 2 + \cdots + n = \frac{n(n+1)}{2}$$

$$1^2 + 2^2 + \cdots + n^2 = \frac{n(n+1)(2n+1)}{6}.$$

$\Longrightarrow$ SOLUTION.

We first prove the first identity. Define a function as follow:

```
g[n_] := n (n + 1)/2
```

Clearly the identity is valid for $n = 1$.

```
g[1]
1
```

Now we suppose it is correct for k and check that it is also valid for $k + 1$. Thus we assume

$$1 + 2 + \cdots k = k(k+1)/2 = g(k),$$

and we need to show that

$$1 + 2 + \cdots + k + k + 1 = g(k) + 1 = g(k+1).$$

We ask *Mathematica* for help

```
Simplify[g[k] + k + 1 == g[k + 1]]
True
```

Thus we proved, by mathematical induction, that the first identity is valid.

Here is the code for the second identity.

```
f[n_] := n (n + 1) (2 n + 1)/6

f[1]
1

f[n] + (n + 1)^2
(1 + n)^2 + 1/6 n (1 + n) (1 + 2 n)

f[n + 1]
1/6 (1 + n) (2 + n) (1 + 2 (1 + n))

Simplify[f[n] + (n + 1)^2 == f[n + 1]]
True
```

There are other ways to define a function (a pattern). For example, we can define

```
x_⊕y_:=x+y+x*y
```

We check this in fact operates correctly.

```
2⊕3= 11
x⊕0= x
a ⊕ b= a + b+ a*b
```

We look at these types of definitions systematically in Problem 10.10 (see also Problem 8.17).

One can define functions with pre-defined (default) values. Here we define the function $f(x) = \sqrt{x}$ so that if we don't pass any value to f, it will return 1.

```
f[x_: 1] := Sqrt[x]

f[16]
4

f[]
1
```

Here is another example, a function with two variables, both given default values. We use two built-in functions Text and Style to change the size and the colour of the text we pass into the function (see Problem 3.7).

```
g[x_: "Freshwater", n_: 10] := Text[Style[x, Red, n]]

g["smell of wine", 30]
```

smell of wine

```
g[]
```

Freshwater

Later, in Chapter 11, we will look at these types of functions further. We will also define functions with conditions, functions with several definitions and functions containing several lines of code (a procedure).

3.2 Anonymous functions

Sometimes we need to "define a function as we go" and use it on the spot. *Mathematica* enables us to define a function without giving it a name (nor any reference to any specific variables), use it, and then move on! These functions are called *anonymous* or *pure* functions. Obviously if we need to use a specific function frequently, then the best way is to give it a name and define it as we did in Section 3.1. Here is an anonymous function equivalent to $f(x) = x^2 + 4$:

```
(#^2+4)&
```

The expression `(#^2+4)&` defines a nameless function. As usual we can plug in data in place of `#`. The symbol `&` determines where the definition of the function is completed.

```
(#^2+4)&[5]
  29
```

Compare the following:

```
r[x_] := x (x + 1)

r[4]
20

# (# + 1) &[4]
20

4 // # (# + 1) &
20
```

Anonymous functions can handle several variables. Here is an example of an anonymous function for $f(x, y) = \sqrt{x^2 + y^2}$.

```
Sqrt[#1^2+#2^2]&[3,4]
  5
```

As you might guess, `#1` and `#2` refer to the first and second variables in the function.

───── **Problem 3.7**

Describe what the following function does.

```
Text[Style["Sydney", Blue, Italic, #]] &
```

⟹ SOLUTION.

We have put several of *Mathematica*'s built-in functions together. The argument (which, here, should be a number) passed to this function will be placed where `#` is. If one looks at the Help for `Style`, one sees that the number will determine the size of the font. Then `Text` will present the word Sydney, in blue italic and the size given in `#`.

```
?Text
Text[expr] displays with expr in plain text format.

?Style
Style[expr,options] displays with expr formatted using the
 specified option settings.
```

Here is the result of applying the function to the number 80.

```
Text[Style["Sydney", Blue, Italic, #]] &[80]
```

Sydney

Exercise 3.2

Investigate what the following pure functions do.

```
Fibonacci[15]//Mod[#,5]&

PrimeQ[#! + 1] &@4

18//2^#+#&

Plot[Sin[x^2], {x, 0, 2 Pi}, PlotStyle -> {#1, #2}] &[Red, Thick]

(x #1 + y #2) &[1, 2]
```

$$\frac{4}{\textit{Lists}}$$

A list is a collection of data. In this chapter we study how to handle lists and have access to the elements in the lists. Functions which generate lists are studied and methods of selecting elements from a list with a specific property are examined. Get ready for some serious programming!

One can think of a computer program as a function which accepts some (crude) data or information and gives back the data we would like to obtain. *Lists* are the way Wolfram *Mathematica*® handles information. Roughly speaking, a list is a collection of objects. The objects could be of any type and pattern. Let us start with an example of a list:

$$\{1, -4/7, \texttt{stuff}, 1-2\,\texttt{x}+\texttt{x}^2, \quad , \texttt{Sin[x^2+y^2]}\}$$

As is shown, the objects (of any type and format) are arranged between brackets, { , }. This looks like a mathematical set. One difference is that lists respect order:

```
{1, 2} == {2, 1}
False
```

The other difference is that a list can contain a copy of the same object several times:

```
{1,2,1} == {1,2}
False
```

© Springer International Publishing Switzerland 2015
R. Hazrat, *Mathematica®: A Problem-Centered Approach*, Springer Undergraduate
Mathematics Series, DOI 10.1007/978-3-319-27585-7_4

The natural thing here is to be able to access the elements of a list. Let us define a list p as follows:

```
p = {x, 1, -8/3, a, b, {c, d, e}, radio}
{x, 1, -(8/3), a, b, {c, d, e}, radio}
```

As this example shows, the list p can have any sort of data, even another list, as one of its elements. Here is how we can access the elements of the list:

```
p[[1]]
x

p[[5]]
b

p[[-1]]
radio

p[[{2, 4}]]
{1, a}

p[[{-2, 5}]]
{{c, d, e}, b}

p[[-2, {2, 3}]]
{d, e}
```

Examining the above examples reveals that p[[i]] gives the i-th member of the list.

To obtain consecutive elements of a list one can use ;;.

```
p = {x, 1, -8/3, a, b, {c, d, e}, radio}

p[[2 ;; 6]]
{1, -(8/3), a, b, {c, d, e}}

p[[3 ;;]]
{-(8/3), a, b, {c, d, e}, radio}

p[[2 ;; 6 ;; 2]]
{1, a, {c, d, e}}
```

Here is an explanation:

```
?;;
i ;; j represents a span of elements i through j.
i ;; represents a span from i to the end.
;; j represents a span from the beginning to j.
;; represent a span that includes all elements.
i ;; j ;; k represents a span from i though j in steps of k.
```

There are other commands to access the elements of a list as follows:

```
p = {x, 1, -8/3, a, b, {c, d, e}, radio}
{x, 1, -(8/3), a, b, {c, d, e}, radio}

First[p]
x

Last[p]
radio

Drop[p, 3]
{a, b, {c, d, e}, radio}

Take[p, 2]
{x, 1}

Rest[p]
{1, -(8/3), a, b, {c, d, e}, radio}

Rest[%]
{-(8/3), a, b, {c, d, e}, radio}

Rest[%]
{a, b, {c, d, e}, radio}

Most[p]
{x, 1, -(8/3), a, b, {c, d, e}}

Most[Rest[p]] == Rest[Most[p]]
True
```

Most of these commands are self-explanatory and a close look at the above examples shows what each of them will do (see the Tips on page 33 for %). All these commands and more are listed in the *Mathematica* Help under Wolfram Documentation: Core Language: List: Elements of Lists.

━━ Problem 4.1

Let p={a,b,{c,d},e}. From this list produce the list {a,b,c,d,e}.

⟹ SOLUTION.

```
{p[[1]], p[[2]], p[[3, 1]], p[[3, 2]], p[[4]]}
{a, b, c, d, e}
```

One can also use the command Flatten to get rid of extra { and }. We will discuss this command in Section 8.2, where we will be looking at nested loops which create nested lists.

```
p = {a, b, c, {d, e}, f}
{a, b, c, {d, e}, f}
```

```
?Flatten
Flatten[list] flattens out nested lists.

Flatten[p]
{a, b, c, d, e, f}
```

Quite often we will be dealing with a list of lists. These occur, for example, when one imports data from spreadsheets. The following example explores how to handle data of this kind.

```
data = {{{{a, b}, 2}, {{c, d}, 3}}, {{{a, b}, 2}, {{e, f}, 3}}}
{{{{a, b}, 2}, {{c, d}, 3}}, {{{a, b}, 2}, {{e, f}, 3}}}

data[[ ;; ]]
{{{{a, b}, 2}, {{c, d}, 3}}, {{{a, b}, 2}, {{e, f}, 3}}}

data[[All]]
{{{{a, b}, 2}, {{c, d}, 3}}, {{{a, b}, 2}, {{e, f}, 3}}}
```

Notice that data contains two elements, each of which is a list itself.

```
data[[1]]
{{{a, b}, 2}, {{c, d}, 3}}

data[[2]]
{{{a, b}, 2}, {{e, f}, 3}}

data[[1, 1]]
{{a, b}, 2}

data[[2, 2]]
{{e, f}, 3}

data[[All, 1]]
{{{a, b}, 2}, {{a, b}, 2}}

data[[All, 2]]
{{{c, d}, 3}, {{e, f}, 3}}

data[[All, All, 1]]
{{{a, b}, {c, d}}, {{a, b}, {e, f}}}
```

There are several commands in *Mathematica* to keep track and access the different layers (or levels) of a list. Level[expr.,{k}] gives the kth layer of the expr., i.e., it returns a list consisting of those expressions which are at level k. However, Level[expr.,k] gives all the layers at or below k-th level. Consider the following list:

```
list = {a, {b, {c, d, e}}}
```

In order to understand the structure of this list (and its layers), we use the command TreeForm to see how this list is composed.

```
?TreeForm

TreeForm[expr] displays expr as a tree with different levels at
different depths.
TreeForm[expr,n] displays expr as a tree only down to level n.>>

TreeForm[list]
```

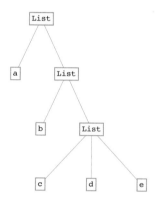

The command Depth gives the depth (number of layers) of this list.

```
Depth[list]
4
```

Note that the head List appearing on the top of the above TreeForm is considered to be the zero level of the expression. Now having the definition of Level in mind, we can justify

```
Level[list, {1}]
{a, {b, {c, d, e}}}

Level[list, {2}]
{b, {c, d, e}}

Level[list, {3}]
{c, d, e}

Level[list, 1]
{a, {b, {c, d, e}}}

Level[list, 2]
{a, b, {c, d, e}, {b, {c, d, e}}}
```

```
Level[list, 3]
{a, b, c, d, e, {c, d, e}, {b, {c, d, e}}}

Level[list, {-2}]
{{c, d, e}}

Level[{{{{x}}}}, {3}]
{{x}}

Level[{{{{x}}}}, 3]
{{x}, {{x}}, {{{x}}}}

Level[{{{{x}}}}, -1]
{x, {x}, {{x}}, {{{x}}}}
```

_____ Problem 4.2

Consider the list

data = {{{{a, b}, 2}, {{c, d}, 3}}, {{{a, b}, 2}, {{e, f}, 3}}}

Using the function Level, from this list produce the following list

{a, b, c, d, a, b, e, f}

⟹ SOLUTION.

The list data comprises of several layers of lists. With the function Level one can get access to each layer (or level) of the list as follows:

```
Level[data, {1}]
{{{{a, b}, 2}, {{c, d}, 3}}, {{{a, b}, 2}, {{e, f}, 3}}}

Level[data, {2}]
{{{a, b}, 2}, {{c, d}, 3}, {{a, b}, 2}, {{e, f}, 3}}

Level[data, {3}]
{{a, b}, 2, {c, d}, 3, {a, b}, 2, {e, f}, 3}

Level[data, {4}]
{a, b, c, d, a, b, e, f}
```

_____ Problem 4.3

Consider the list a

```
a={{x[1, 1], x[1, 2], x[1, 3]}, {x[2, 1], x[2, 2], x[2, 3]},
   {x[3, 1], x[3, 2], x[3, 3]}}
```

This is a list consisting of two layers. One can consider this as a 3×3 matrix as follows

```
a // MatrixForm
```

$$\begin{pmatrix} x[1,1] & x[1,2] & x[1,3] \\ x[2,1] & x[2,2] & x[2,3] \\ x[3,1] & x[3,2] & x[3,3] \end{pmatrix}$$

From this matrix, extract its boundary entries.

$\Longrightarrow$ SOLUTION.

The list a is generated by the command Array.

```
a = Array[x, {3, 3}]
```

We will study this command and others that generate matrices in Chapter 13.

The boundaries can be obtained by using All to get all the elements in the different layers.

```
a[[All, 1]]
{x[1, 1], x[2, 1], x[3, 1]}

a[[1, All]]
{x[1, 1], x[1, 2], x[1, 3]}

a[[All, -1]]
{x[1, 3], x[2, 3], x[3, 3]}

a[[-1, All]]
{x[3, 1], x[3, 2], x[3, 3]}
```

One of the secrets of writing code comfortably is that one should be able to manipulate lists easily. Often in applications, situations like the following arise:

- Given $\{x_1, x_2, \cdots, x_n\}$ and $\{y_1, y_2, \cdots, y_n\}$, produce

$$\{x_1, y_1, x_2, y_2, \cdots, x_n, y_n\},$$

and

$$\{\{x_1, y_1\}, \{x_2, y_2\}, \cdots, \{x_n, y_n\}\}.$$

- Given $\{x_1, x_2, \cdots, x_n\}$ and $\{y_1, y_2, \cdots, y_n\}$, produce

$$\{x_1 + y_1, x_2 + y_2, \cdots, x_n + y_n\}.$$

– Given $\{x_1, x_2, \cdots, x_n\}$ and $\{y_1, y_2, \cdots, y_n\}$, produce

$$\{\{x_1, y_1\}, \{x_1, y_2\}, \cdots, \{x_1, y_n\},$$
$$\{x_2, y_1\}, \{x_2, y_2\}, \cdots, \{x_2, y_n\}, \cdots, \{x_n, y_1\}, \{x_n, y_2\}, \cdots, \{x_n, y_n\}\}.$$

– Given $\{x_1, x_2, \cdots, x_n\}$, produce

$$\{x_1, x_1 + x_2, \cdots, x_1 + x_2 + \cdots + x_n\}.$$

– Given $\{x_1, x_2, \cdots, x_n\}$, produce

$$\Big\{\{\{x_1\}, \{x_2, \ldots, x_n\}\}, \{\{x_1, x_2\}, \{x_3, \ldots, x_n\}\} \cdots$$
$$\{\{x_1, \ldots x_{n-1}\}, \{x_n\}\}\Big\}.$$

Mathematica provides us with commands to obtain the above arrangements easily. We will look at these commands in Sections 8.4, 8.5, 13.1 and Problem 10.7.

♣ TIPS

– One can add elements to a list. There are several commands to handle this, including `Append`, `Prepend` and `Insert`.

```
?Insert

Insert[list,elem,n] insert elem at position n in list. If n is
negative, the position is counted from the end.

Insert[{i, think, i, am}, "therefore", 3]
{i, think, therefore, i, am}
```

See also Example 6.3 for the use of `Append`.

– The commands `Sort`, `Reverse`, `RotateLeft` and `RotateRight` are available to rearrange the order of a list.

– The commands `Delete` and `Drop` are available to remove elements from a list.

4.1 Functions producing lists

Mathematica provides us with commands for which the output is a list. These commands have a nature of repetition and replace loops in procedural programming (more on this in Chapter 8). Let us look at some of them here before starting to write more serious codes.

```
Range[10]
{1, 2, 3, 4, 5, 6, 7, 8, 9, 10}

Range[3, 11]
{3, 4, 5, 6, 7, 8, 9, 10, 11}

Range[2, 17, 4]
{2, 6, 10, 14}

Range[-12, 3, 2]
{-12, -10, -8, -6, -4, -2, 0, 2}

Range[0.25, 1, 0.1]
{0.25, 0.35, 0.45, 0.55, 0.65, 0.75, 0.85, 0.95}

?Range

Range[imax] generates the list {1,2,...,imax}.
Range[imin, imax] generates the list {imin,...,imax}.
Range[imin,imax,di] uses step di.>>
```

One of the most useful commands which produces a list is `Table`.

```
Table[2 n + 1, {n, 1, 13}]
{3, 5, 7, 9, 11, 13, 15, 17, 19, 21, 23, 25, 27}
```

In `Table[2n+1,{n,1,13}]`, n runs from 1 to 13 and each time the *function* 2n+1 is evaluated.

The following example shows how easily we can work symbolically in *Mathematica*.

```
Table[x^i + y^i, {i, 2, 17, 4}]
{x^2 + y^2, x^6 + y^6, x^10 +y^10, x^14 + y^14}
```

As in the last example of `Range`, here in `Table`, i starts from 2 with step 4 and thus takes the values 2,6,10,14.

Here is one more example demonstrating how beautifully *Mathematica* can handle symbols

```
Table[x_i, {i,1,10}]
```
$$\{x_1, x_2, x_3, x_4, x_5, x_6, x_7, x_8, x_9, x_{10}\}$$

One can evaluate any kind of function (expression) in the `Table`:

```
f[i_] := Plot[Sin[x^i], {x, 0, Pi}]

Table[f[i], {i, 1, 4}]
```

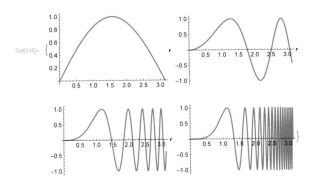

So far `Table` has used a range of numbers. In fact we can assign a list as the range for the `Table`.

> `Table[expr,{i, {i1,i2,...}]` uses the successive values i1,i2,...

As an example, we assign the list of the four plots above to the variable `glist`. Then we ask `Table` to run through this list and rotate each of these plots:

```
Table[Rotate[i, Random[]], {i, glist}]
```

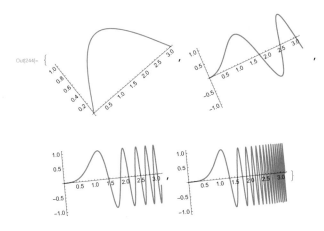

Problem 4.4

Produce the list of the first 30 Fibonacci numbers.

⟹ SOLUTION.

All we have to do is to let i run from 1 to 30 and each time plug i into the function Fibonacci[i] which produces the i-th Fibonacci number. This can be done with Table as follows:

```
Table[Fibonacci[i], {i, 1, 30}]
{1, 1, 2, 3, 5, 8, 13, 21, 34, 55, 89, 144, 233, 377, 610,
987, 1597, 2584, 4181, 6765, 10946, 17711, 28657, 46368, 75025,
121393, 196418, 317811, 514229, 832040}
```

Problem 4.5

Find all natural numbers n between 1 and 12 for which the polynomial $x^n + 64$ can be written as a product of two nonconstant polynomials with integer coefficients.

⟹ SOLUTION.

We have seen that Factor will give a factorization of an expression (if it is possible).

```
Table[Factor[x^i + 64], {i, 1, 10}]
{64 + x, 64 + x^2, (4 + x) (16 - 4 x + x^2),
(8 - 4 x + x^2) (8 + 4 x + x^2),   64 + x^5,
(4 + x^2) (16 - 4 x^2 + x^4),   64 + x^7,
(8 - 4 x^2 + x^4) (8 + 4 x^2 + x^4),
(4 + x^3) (16 - 4 x^3 + x^6), 64 + x^10}
```

Although this solves the problem (for $1 \leq n \leq 10$), the output is not arranged nicely and it is hard to understand the answer. It would be easier if we keep track of i as well. We change the code slightly. For this we need Print

```
?Print
Print[expr] prints expr as output. >>
```

Here is the improved code:

```
Table[Print[i, "   ", Factor[x^i + 64]], {i, 1, 12}];

1   64+x

2   64+x^2
```

3   (4+x) (16-4 x+x^2)

4   (8-4 x+x^2) (8+4 x+x^2)
5   64+x^5

6   (4+x^2) (16-4 x^2+x^4)

7   64+x^7

8   (8-4 x^2+x^4) (8+4 x^2+x^4)

9   (4+x^3) (16-4 x^3+x^6)

10   64+x^10

11   64+x^11

12   (2-2 x+x^2) (2+2 x+x^2) (4-4 x+2 x^2-2 x^3+x^4)
     (4+4 x+2 x^2+2 x^3+x^4)

In terms of presentation, one could do a bit better, for example try,

```
Table[Print[x^i + 64, " = ", Factor[x^i + 64]], {i, 1, 15}];
```

From the list, for $n = 3, 4, 6, 8, 9, 12$ we have a factorisation into two non-constant polynomials (in fact, for $n = 3k$ and $4k$ the polynomial can be factorised into two polynomials).

▌━━━━━━━━━━━━━━━━

━━━━ Problem 4.6

Determine all the positive integers n between 3 and 50 for which 2^{2008} is divisible by

$$1 + \binom{n}{1} + \binom{n}{2} + \binom{n}{3}.$$

⟹ SOLUTION.
   Recall that if Mod[m,n] returns zero, then m is divisible by n. We first define a function $b(n) = 1 + \binom{n}{1} + \binom{n}{2} + \binom{n}{3}$ (see Problem 3.4) and then use a Table to check when 2^{2008} is divisible by $b(n)$ for different n.

```
b[n_] := Binomial[n, 1] + Binomial[n, 2] + Binomial[n, 3] + 1

Table[Mod[2^2008, b[n]], {n, 3, 50}]
```

{0, 1, 16, 16, 0, 70, 16, 80, 168, 133, 268, 316, 448, 256,
706, 796, 1096, 723, 1092, 1030, 0, 256, 458, 1240, 2704, 2604,
606, 2922, 640, 1664, 2704, 4076, 2824, 1936, 1024, 8336, 256,
7882, 6974, 4192, 4568, 6061, 1076, 6896, 704, 16669, 6856,
16032}

It is clear from the list that only for $n = 3, 7$ and 23 is 2^{2008} divisible by the above expression (in fact, one can show that these are the only numbers with this property).

Here is a nice example showing the difference between the two equalities $=$ and $:=$.

Example 4.7

This example uses `BarChart` which is a graphic function. The example is based on the discussion in Section 2.6.

```
?BarChart
BarChart[{y1,y2,...}] makes a bar chart with bar lengths y1,y2,...

x=RandomInteger[{1,1000}];
BarChart[Table[x,{200}]]
```

Figure 4.1   Using $=$ as equality

```
x:=RandomInteger[{1,1000}]
BarChart[Table[x,{200}]]
```

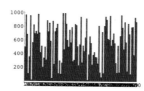

Figure 4.2   Using $:=$ as equality

In order to understand this example better, get the list generated by `Table[x,{1000}]` for each of the definitions of x individually and compare them.

4.2 Listable functions

There are times when we would like to apply a function to all the elements
of a list. Suppose f is a function and {a,b,c} is a list. We want to be able
to "push" the function f inside the list and get {f[a],f[b],f[c]}. Many of
Mathematica's built-in functions have the property that they simply "go inside"
a list. A function with this property is said to be *listable*. For example Sqrt is
a listable function.

```
Sqrt[{a, b, c, d, e}]
```

$$\{\sqrt{a}, \sqrt{b}, \sqrt{c}, \sqrt{d}, \sqrt{e}\}$$

All the arithmetic functions are listable:

```
1+ {a, b, c, d, e}
{1 + a, 1 + b, 1 + c, 1 + d, 1 + e}

{a, b, c, d, e}^3
{a^3, b^3, c^3, d^3, e^3}

1/{a, b, c, d, e}
{1/a, 1/b, 1/c, 1/d, 1/e}
```

We will use the function Sqrt in the following to show that the product of
four consecutive numbers plus one is always a square number (for a proof of
this statement, see page 38).

```
Table[n (n + 1) (n + 2) (n + 3) + 1, {n, 1, 10}]
{25, 121, 361, 841, 1681, 3025, 5041, 7921, 11881, 17161}

Sqrt[%]
{5, 11, 19, 29, 41, 55, 71, 89, 109, 131}
```

Not all functions are listable. If we want to push a function into a list, the
command to use is Map.

```
f[{a, b, c, d, e}]
f[{a, b, c, d, e}]

Map[f, {a, b, c, d, e}]
{f[a], f[b], f[c], f[d], f[e]}
```

The equivalent shorthand to apply a function to a list is /@ as follows:

```
f /@ {a, b, c, d, e}
{f[a], f[b], f[c], f[d], f[e]}

s/@ Range[10]
{s[1], s[2], s[3], s[4], s[5], s[6], s[7], s[8], s[9], s[10]}
```

Here are some more examples:

```
Table[(1 + x)^i, {i, 5}]
{1 + x, (1 + x)^2, (1 + x)^3, (1 + x)^4, (1 + x)^5}

Expand /@ %
{1 + x, 1 + 2 x + x^2, 1 + 3 x + 3 x^2 + x^3,
 1 + 4 x + 6 x^2 + 4 x^3 + x^4,
 1 + 5 x + 10 x^2 + 10 x^3 + 5 x^4 + x^5}

Range[0.25, 1, 0.1] == Range[2.5, 10, 1]/10
True

?Framed
Framed[expr] displays a framed version of expr.

?Nest
Nest[f,expr,n] gives an expression with f applied n times to expr

f[x_] := 1/(1 + x)

Framed@ Nest[f, x, #] & /@  Range[10]
```

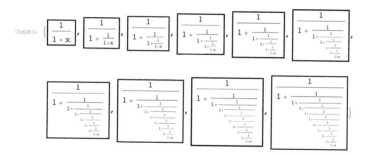

Here is another example, showing practically anything can be seen as a function and thus sent into a list. Here our pure function is

```
Plot[Sin[1/x],{x,-Pi,Pi},PlotStyle->#] &
```

and our list is {Red, Blue, Green, Yellow}.

In[126]:= `c = {Red, Blue, Green, Yellow}`

Out[126]= $\{\blacksquare, \blacksquare, \blacksquare, \square\}$

In[128]:= `Plot[Sin[1 / x], {x, -Pi, Pi}, PlotStyle → #] & /@ c`

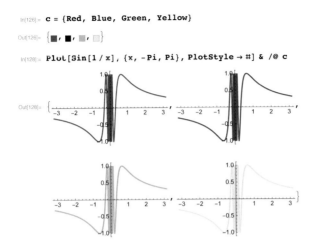

Out[128]=

Problem 4.8

The formula $n^2 + n + 41$ has a very interesting property. Observe that this formula produces prime numbers for all n between 0 and 39.

⟹ SOLUTION. First we produce the numbers:

```
Table[n^2 + n + 41, {n, 0, 40}]
{41, 43, 47, 53, 61, 71, 83, 97, 113, 131, 151, 173, 197,
223, 251, 281, 313, 347, 383, 421, 461, 503, 547, 593, 641, 691,
743, 797, 853, 911, 971, 1033, 1097, 1163, 1231, 1301, 1373,
1447, 1523, 1601, 1681}
```

Then we apply PrimeQ to this list. This function is listable.

```
PrimeQ[%]
{True, True, True, True, True, True, True, True, True, True,
True, True, True, True, True, True, True, True, True, True,
True, True, True, True, True, True, True, True, True, True,
True, True, True, True, True, True, True, True, False}
```

The list shows that for n between 0 and 39, the above formula produces prime numbers, however, for $n = 40$, the last number in the list $40^2 + 40 + 41$ is not prime. This formula was found by Euler around 1772. One wonders, if one changes 41 to another number in the formula $n^2 + n + 41$, whether one gets more consecutive prime numbers. We will examine this in Problem 8.2.

Of course we could also simply write

```
Table[PrimeQ[n^2 + n + 41], {n, 0, 40}]
```

as well. As we said, there are many ways to approach a problem in *Mathematica*.

━━━━━━━━

━━━━ Problem 4.9

Find the number of all the words starting with "a", "b", ..., and "z" available in *Mathematica*'s dictionary.

⟹ SOLUTION.

We need some of *Mathematica*'s built-in functions to put this code together. First `CharacterRange` can produce the list of all the alphabetical characters from "a" to "z".

```
CharacterRange["a", "z"]
{"a", "b", "c", "d", "e", "f", "g", "h", "i", "j", "k", "l", "m",
 "n", "o", "p", "q", "r", "s", "t", "u", "v", "w", "x", "y", "z"}
```

We need *Mathematica*'s dictionary, which is called `DictionaryLookup`.

```
DictionaryLookup[patt] finds all words in an English dictionary
    that match the string pattern patt
```

Finally, for all the words starting with "a", we write a~~___. We will discuss the role of blanks ___ in Chapter 10 when we deal with patterns. For the moment, we can think of this line as describing all the words starting with "a". Now in the code below, we replace "a" with all the other letters using the pure function #~~___ and then apply it to the list of all letters.

```
Length[DictionaryLookup[# ~~ ___]] & /@ CharacterRange["a", "z"]

{4500, 4724, 7831, 5198, 3244, 3446, 2626, 2954, 3357, 711,
 577, 2392, 4196, 1698, 2104, 6559, 412, 5143, 9599, 4160, 2561,
 1206, 2205, 19, 248, 137}
```

Finally, we leave it to the reader to decipher what the following enhanced code would do.

```
Print[#, " ", Length[DictionaryLookup[# ~~ ___]]] & /@
    CharacterRange["a", "z"];
```

━━━━━━━━

Exercise 4.1

Generate 6 random integers n between 1 and 3095 and show that $n^6 + 1091$ is not prime. (Hint: see `RandomInteger` for generating random numbers.)

Exercise 4.2

Describe what the following line does (see Problem 3.7):

```
Text[Style["Sydney", Red, Italic, #]] & /@ Range[40, 70, 10]
```

4.3 Selecting from a list

So far we have been able to create a list of data by using functions producing lists. The next step is to be able to choose, from a list, certain data which fit a specific description. This can be achieved using the command `Select`, as the following problem demonstrates.

▬▬ Problem 4.10

How many numbers of the form $3n^5 + 11$, when n varies from 1 to 2000, are prime?

⟹ SOLUTION.

First, let us produce the first 20 numbers of this form.

```
plist=Table[3 n^5 + 11, {n, 1, 20}]
{14, 107, 740, 3083, 9386, 23339, 50432, 98315, 177158,
300011, 483164, 746507, 1113890, 1613483, 2278136, 3145739,
4259582, 5668715, 7428308, 9600011}
```

The next step, in the spirit of Section 4.2, would be to apply `PrimeQ` to all the numbers and find out which ones are prime. Since this is a listable function it is enough to do the following:

```
PrimeQ[plist]
{False, True, False, True, False, True, False, False, False,
False, False, True, False, True, False, True, False, False,
False, False}
```

Now we are left to count the number of `True` responses in the list. This is easily done here; 6 of these numbers are prime. However, if we are dealing with a list with 2000 elements, we need to select, from the elements of the list, those with a desired property,[1] here being the elements which are prime. The command `Select` does just this:

```
Select[plist,PrimeQ]
{107, 3083, 23339, 746507, 1613483, 3145739}
```

---
[1] Or a desired pattern, more about this later.

These are prime numbers in the list `plist`. The command `Length` gives the length of a list. If we assemble all the steps in one line we have

```
Length[Select[Table[3 n^5 + 11, {n, 1, 20}], PrimeQ]]
6
```

Thus to find out how many numbers of the form $3n^5 + 11$ are prime when n runs from 1 to 2000, all we have to do is to change 20 to 2000:

```
Length[Select[Table[3 n^5 + 11, {n, 1, 2000}], PrimeQ]]
97
```

Here is another approach to this problem. We produce a list of True and False by applying `PrimeQ` to the numbers we generate and then count the number of Trues using the command `Count`.

```
?Count

Count[list,pattern] gives the number of elements in list that
match pattern

Table[PrimeQ[3 n^5 + 11], {n, 1, 10}]
{False, True, False, True, False, True, False, False, False,
False}

Count[Table[PrimeQ[3 n^5 + 11], {n, 1, 10}], True]
3
```

We now run the code for the first 2000 numbers generated by $3n^5 + 11$.

```
Count[Table[PrimeQ[3 n^5 + 11], {n, 1, 2000}], True]
97
```

In a nutshell, `Select[list,f]` will apply the function `f` (which returns `True` or `False`) to all the elements, say x, of the `list` and return those elements for which `f[x]` is true. A function which returns either `True` or `False` is called a *Boolean* function. In mathematical syntax, this is

$$\{x \in list \mid f(x)\}.$$

So far we have seen one Boolean function, that is, `PrimeQ`. (for more on Boolean statements, see Chapter 6.)

——— Problem 4.11

Show that among the first 500 Fibonacci numbers, 18 of them are prime.

$\Longrightarrow$ SOLUTION.

We do this step by step. First we generate the first 10 Fibonacci numbers. Then among those we select the ones which are prime. Then using Length we get the size of this list, as follows:

```
fiblist=Table[Fibonacci[i], {i, 1, 10}]
{1, 1, 2, 3, 5, 8, 13, 21, 34, 55}

Select[fiblist, PrimeQ]
   {2, 3, 5, 13}

Length[Select[Table[Fibonacci[i], {i, 1, 10}], PrimeQ]]
   4
```

Now instead of 10 we generate 500 numbers:

```
Length[Select[Table[Fibonacci[i], {i, 1, 500}], PrimeQ]]
   18
```

Exercise 4.3

Show that, among the first 450 Fibonacci numbers, the number of odd Fibonacci numbers is twice the number of even ones. (Hint: see OddQ and EvenQ for odd and even numbers.)

Exercise 4.4

Let m be a natural number and

$$A = \frac{(m+3)^3 + 1}{3m}.$$

Find all the integers m less than 500 such that A is an integer. Show that A is always odd. (Hint: see IntegerQ.)

Let us look at another example, slightly different but of the same nature as Problem 4.10. The following example shows that anonymous functions fit very well with Select.

Problem 4.12

For which $1 \leq n \leq 1000$ does the formula $2^n + 1$ produce a prime number?

$\Longrightarrow$ SOLUTION. Here is the solution:

```
Select[Range[1000], PrimeQ[2^(#) + 1] &]
   {1, 2, 4, 8, 16}
```

Let us take a deep breath and go through this one-liner slowly. The function `PrimeQ[2^(#)+1]&` is an anonymous function which returns `True` if the number $2^n + 1$ is prime and `False` otherwise.

```
PrimeQ[2^(#) + 1] &[12]
False
```

`Range[1000]` creates a list containing the numbers from 1 to 1000. The command `Select` applies the anonymous function `PrimeQ[2^(#)+1]&` to each element of this list and in the cases when the result is true, i.e., when $2^n + 1$ is prime, the element will be selected. Thus {1, 2, 4, 8, 16} are the only numbers n in the given range for which $2^n + 1$ is a prime number.

Again, *Mathematica* allows us to use different approaches to solve the problem. We provide two more solutions to this problem. The reader can choose their favourite solution or come up with yet another one!

The first code below uses the built-in function `Position`. We generate a list of True and False based on whether the number $2^n + 1$ is prime or not, and then find the position of each True which correspond to the n such that $2^n + 1$ is prime.

```
?Position
Position[expr,pattern] gives a list of the positions at which
objects matching pattern appear in expr.

Position[Table[PrimeQ[2^i + 1], {i, 1, 1000}], True]
{{1}, {2}, {4}, {8}, {16}}
```

The next approach uses the built-in function `Pick`.

```
?Pick

Pick[list,sel] picks out those elements of list for which
the corresponding element of sel is True.

Pick[Range[1000], Table[PrimeQ[2^i + 1], {i, 1, 1000}]]
{1, 2, 4, 8, 16}
```

Exercise 4.5

Find the number of positive integers $0 < n < 20000$ such that 1997 divides $n^2 + (n + 1)^2$. Try the same code for 2009.

Exercise 4.6

For integers $2 \leq n \leq 200$, find all n such that n divides $(n - 1)! + 1$. Show that there are 46 such n.

_____ Problem 4.13

Notice that $12^2 = 144$ and $21^2 = 441$, i.e., the numbers and their squares are reverses of each other. Find all the numbers up to 10000 with this property.

$\Longrightarrow$ SOLUTION.

We need to introduce some new built-in functions. `IntegerDigits[n]` gives a list of the decimal digits of the integer n. We also need `Reverse` and `FromDigits`:

```
IntegerDigits[80972]
{8, 0, 9, 7, 2}

Reverse[%]
{2, 7, 9, 0, 8}

FromDigits[%]
27908
```

Thus the above shows we can easily produce the reverse of a number:

```
re[n_] := FromDigits[Reverse[IntegerDigits[n]]]

re[12345]
54321

re[6548629]
9268456
```

Having this function under our belt, the solution to the problem is just one line. Notice that the problem is asking for the numbers n such that re[n^2]=re[n]^2 .

```
Select[Range[10000], re[#]^2 == re[#^2] &]
{1, 2, 3, 10, 11, 12, 13, 20, 21, 22, 30, 31, 100, 101,
102, 103, 110, 111, 112, 113, 120, 121, 122, 130, 200, 201, 202,
210, 211, 212, 220, 221, 300, 301, 310, 311, 1000, 1001, 1002,
1003, 1010, 1011, 1012, 1013, 1020, 1021, 1022, 1030, 1031, 1100,
1101, 1102, 1103, 1110, 1111, 1112, 1113, 1120, 1121, 1122, 1130,
1200, 1201, 1202, 1210, 1211, 1212, 1220, 1300, 1301, 2000, 2001,
2002, 2010, 2011, 2012, 2020, 2021, 2022, 2100, 2101, 2102, 2110,
2111, 2120, 2121, 2200, 2201, 2202, 2210, 2211, 3000, 3001, 3010,
3011, 3100, 3101, 3110, 3111, 10000}
```

Here is one more example using the command `FromDigits`. We know that 11 is a prime number. One wonders what is the next prime number consisting only of ones. A wild guess: a number with 23 digits all one? All we have to do is to produce this number then, with `PrimeQ`, test whether this is prime. Here is one way to generate this number.

```
Table[1, {23}]
{1, 1, 1, 1, 1, 1, 1, 1, 1, 1, 1, 1, 1, 1, 1, 1, 1,
 1, 1, 1, 1, 1, 1}

FromDigits[%]
11111111111111111111111

PrimeQ[%]
True
```

Here is the code to find out which numbers of this kind up to 500 digits are prime.

```
Select[Range[500], PrimeQ[FromDigits[Table[1, {#}]]] &]
{2, 19, 23, 317}
```

────── Problem 4.14

Show that the number of k between 0 and 1000 for which $\binom{1000}{k}$ is odd is a power of 2.

⟹ SOLUTION.

First, Range[0,1000] creates a list of numbers from 0 to 1000. Using an anonymous function, each time we plug these numbers into Binomial[1000, #] and check right away if this is an odd number by OddQ[Binomial[1000, #]] &. If this is the case, Select will pick up these numbers. The rest is clear from the code below.

```
Select[Range[0, 1000], OddQ[Binomial[1000, #]] &]
{0, 8, 32, 40, 64, 72, 96, 104, 128, 136, 160, 168, 192,
 200, 224, 232, 256, 264, 288, 296, 320, 328, 352, 360, 384, 392,
 416, 424, 448, 456, 480, 488, 512, 520, 544, 552, 576, 584, 608,
 616, 640, 648, 672, 680, 704, 712, 736, 744, 768, 776, 800, 808,
 832, 840, 864, 872, 896, 904, 928, 936, 960, 968, 992, 1000}

Length[Select[Range[0, 1000], OddQ[Binomial[1000, #]] &]]
64

FactorInteger[%]
{{2, 6}}
```

Exercise 4.7

Show that among the first 200 primes p, the ones such that the remainder when 19^{p-1} is divided by p^2 is 1 are $\{3, 7, 13, 43, 137\}$.

Exercise 4.8

An integer $d_n d_{n-1} \ldots d_1$ is called *prime-palindromic* if

$$d_n d_{n-1} \ldots d_1 \text{ and } d_1 \ldots d_{n-1} d_n$$

are both prime (for example 941). Write a code to find all prime-palindromic numbers up to 5000. Observe that there are 167 such numbers.

Problem 4.15

Find all positive integers $0 < m < 10^5$ such that the fourth power of the number of positive divisors of m equals m. (Hint: see `Divisors`.)

$\implies$ SOLUTION.

The function `Divisors[n]` gives all the positive numbers which divide n. This is a list which also includes 1 and n. We need to get the number of these divisors, thus we need the function `Length`. We are looking for each number m such that the fourth power of `Length[Divisors[m]]` is m. The rest is clear from the code.

```
Select[Range[10^5], Length[Divisors[#]]^4 == # &]
{1, 625, 6561}
```

Exercise 4.9

Show that there is only one positive integer n smaller than 1000 such that $n! + (n+1)!$ is the square of an integer.

The idea of sending a function into a list, i.e., applying a function to each element of a list, seems to be a good one. We have already mentioned that the listable built-in functions are able to go inside a list, like `PrimeQ` or `Prime`. Have a look at the Attributes of `Prime` in the following:

```
??Prime
Prime[n] gives the nth prime number.
Attributes[Prime] = {Listable, Protected}
```

Also recall that if a function is not listable, `Map` or `/@` will push the function into the list as the following problem demonstrates:

Problem 4.16

What digit does not appear as the last digit of the first 20 Fibonacci numbers?

$\Longrightarrow$ SOLUTION.

This one-liner code collects all the digits which appears as the last digit:

```
Union[Last /@ (IntegerDigits /@ (Fibonacci /@ Range[20]))]
{0, 1, 2, 3, 4, 5, 7, 8, 9}
```

Thus 6 is the only digit which is not present. Let us understand this code. As the command /@ applies a function to all elements of a list, `Fibonacci /@ Range[20]` produces the first 20 Fibonacci numbers.

```
Fibonacci /@ Range[20]
{1, 1, 2, 3, 5, 8, 13, 21, 34, 55, 89, 144, 233, 377, 610,
987, 1597, 2584, 4181, 6765}
```

Then `IntegerDigits` would go inside this list and get the digits of each number.

```
IntegerDigits/@ Fibonacci /@ Range[20]
{{1}, {1}, {2}, {3}, {5}, {8}, {1, 3}, {2, 1}, {3, 4}, {5, 5}, {8,
9}, {1, 4, 4}, {2, 3, 3}, {3, 7, 7}, {6, 1, 0}, {9, 8, 7},
{1, 5, 9, 7}, {2, 5, 8, 4}, {4, 1, 8, 1}, {6, 7, 6, 5}}
```

Then the function `Last` will get the last digits as required.

```
Last /@ IntegerDigits /@ Fibonacci /@ Range[20]
{1, 1, 2, 3, 5, 8, 3, 1, 4, 5, 9, 4, 3, 7, 0, 7, 7, 4, 1, 5}
```

`Union` will get rid of any repetitions in the list. (More on this command in Section 6.2.)

Since `Fibonacci` and `IntegerDigits` are listable functions, one can also write the above code as follows:

```
Union[Last /@ IntegerDigits[Fibonacci[Range[20]]]]
```

If one does not want to use `IntegerDigits` then one can use the `Mod` function to get access to the last digit of a number.

```
?Mod
Mod[m, n] gives the remainder on division of m by n.

Mod[264,10]
4

?Quotient
Quotient[m, n] gives the integer quotient of m and n.

Quotient[264,10]
26

26*10+4
264
```

Thus another way to write the code is as follows. Note that `Mod` is also a listable function.

```
Union[Mod[Fibonacci[Range[20]],10]]
{0, 1, 2, 3, 4, 5, 7, 8, 9}
```

In Section 6.2 we will see how to use *Mathematica* to directly find the digit 6 that we are after, that is how to handle sets.

Recall that one can decompose any number n as a product of powers of primes and that this decomposition is unique, i.e., $n = p_1^{k_1} \cdots p_t^{k_t}$, where the p_i's are prime. Let us call a number a *square free* number if, in its prime decomposition, all the k_i's are 1. That is, no power of primes can divide this number. Thus $15 = 3 \times 5$ is a square free number but $48 = 2^4 \times 3$ is not.

Recall that Select[list,f] will apply the function f (which returns True or False) to all the elements, say x, of the list and return those elements for which f[x] is true. There is an option in Select which makes it possible to get only the first n elements of list that satisfy f, i.e., f returns True, as follows: Select[list,f,n]. This comes in very handy, as in some problems we are only interested in a certain number of data which satisfy f. This can also be used in circumstances where we want to test the elements until something goes wrong or some desirable element comes up. The following example demonstrates this.

Problem 4.17

Write a function squareFreeQ[n] that returns True if the number n is a square free number, and False otherwise.

$\Longrightarrow$ SOLUTION.
    Here is the code:

```
squareFreeQ[n_]:= Select[Last/@ FactorInteger[n], # != 1&,1]
== {}

squareFreeQ[2*5*7]
True

squareFreeQ[2*5*7*7]
False
```

Let us decipher this code. Recall from Section 2.2 that if $n = p_1^{k_1} \cdots p_t^{k_t}$ is the decomposition of n into its prime factors then

```
FactorInteger[n]
```

$$\{\{p_1, k_1\}, \{p_2, k_2\}, \cdots, \{p_t, k_t\}\}$$

Now all we have to do is to go through this list and see if all the k_i's are one. So the first step is to apply Last to each list to discard the p_i's and leave the k_i's.

```
Last /@ FactorInteger[n]
```

$$\{k_1, k_2, \cdots, k_t\}$$

Having obtained this list, we shall go through its elements one by one and examine if the k_i's are one. The anonymous function `# != 1&` does exactly this. Here `!` is negation and thus `# != 1&` means `# ≠ 1&`. So `Select[{k₁,k₂,⋯,kₜ},# != 1&]` gives the list of k_i's which are not one. But in our case, looking for square free primes, it is enough if only one k_i is not one. Then the number is not square free. Thus we use an option of Select which goes through the list until it finds an element such that k_i is not one. So we need to modify the code to `Select[{k₁,k₂,⋯,kₜ},# != 1&,1]`. We are almost done. All we have to do is to see if this list is empty or not (i.e., is there any k_i not equal to one). For this we compare
   `Select[{k₁,k₂,⋯,kₜ},# != 1&,1]=={}`.
We will solve this problem later using slightly different methods (see Problem 5.4 and Problem 10.4).

─────── Problem 4.18

Find the number of integers $0 \le n \le 10^4$ such that a permutation of its digits yield an integer divisible by 11.

⟹ SOLUTION.
We start with an example. Consider the number 134. Using IntegerDigits we get all the digits. We then use the built-in function Permutations to get all the possible permutation of this set. We put these digits back together by mapping FromDigits into the list.

```
IntegerDigits[134]
{1, 3, 4}

?Permutations
Permutations[list] generates a list of all possible permutations
of the elements in list.
```

```
Permutations[IntegerDigits[134]]
{{1, 3, 4}, {1, 4, 3}, {3, 1, 4}, {3, 4, 1}, {4, 1, 3}, {4, 3,1}}

FromDigits /@ Permutations[IntegerDigits[134]]
{134, 143, 314, 341, 413, 431}
```

Next we find among this list which of the numbers are divisible by 11.

```
Select[FromDigits /@ Permutations[IntegerDigits[134]],
  Divisible[#, 11] &]
```

```
{143, 341}
```

For our purpose, we only need to know if there is one permutation which is divisible by 11. Thus we search until such a permutation is found.

```
Select[FromDigits /@ Permutations[IntegerDigits[134]],
  Divisible[#, 11] &, 1]
```

```
{143}
```

We are in a position to put all of this together and define a function.

```
s[n_] := Select[FromDigits /@ Permutations[IntegerDigits[n]],
  Divisible[#, 11] &, 1]
```

```
Length[Select[Range[0, 10^4], s[#] != {} &]]
```

```
2432
```

As always, there are other ways to write the code, such as the following:

```
ss[n_] :=
  Select[FromDigits /@ Permutations[IntegerDigits[n]]/11,
    IntegerQ, 1]
```

Exercise 4.10

Find the first 5 positive integers n such that $n^6 + 1091$ is prime. Show that all these n are between 3500 and 8500.

Exercise 4.11

A number with n digits is called *cyclic* if multiplication by $1, 2, 3, \cdots, n$ produces the same digits in a different order. Find the only 6-digit cyclic number.

Problem 4.19

Find out how many primes bigger than n and smaller than $2n$ exist, when n goes from 1 to 30.

$\Longrightarrow$ Solution.

We define an anonymous function which finds all the primes bigger than n and smaller than $2n$ and then returns the size of this list. Once we are done with this, we apply this function to a list of numbers from 1 to 30. Our anonymous function looks like this:

```
Length[Select[Range[# + 1, 2 # - 1], PrimeQ]] &
```

Here, `Range[# + 1, 2 # - 1]` produces all the numbers between n and $2n$. Then `Select` finds out which of them are in fact prime. Then we use the command `Length` to get the number of elements of this list. One can optimise this a bit, as we don't need to look at the whole range of n to $2n$, as clearly even numbers greater than 2 are not prime so we can ignore them right from the beginning. But we leave it to the reader to do this. All we have to do now is to apply this function with `Map` or `/@` to numbers from 1 to 30.

```
Length[Select[Range[# + 1, 2 # - 1], PrimeQ]] & /@ Range[30]
{0, 1, 1, 2, 1, 2, 2, 2, 3, 4, 3, 4, 3, 3, 4, 5, 4, 4, 4, 4, 5,
6, 5, 6, 6,  6, 7, 7, 6, 7}
```

One can also write a more innocent code, using `Table` as follows:

```
Table[Length[Select[Range[n + 1, 2n - 1], PrimeQ]], {n, 1, 30}]
{0, 1, 1, 2, 1, 2, 2, 2, 3, 4, 3, 4, 3, 3, 4, 5, 4, 4, 4, 4,
5, 6, 5, 6, 6, 6, 7, 7, 6, 7}
```

This doesn't seem to be the most efficient way to solve this problem as each time we are testing the same numbers repeatedly to see whether they are prime. We will write another code for this problem in Problem 8.1 using a Do Loop. One can also use the built-in function `PrimePi` to write a very short solution to this problem. For the usage of `PrimePi`, see Problem 2.4.

A word is called *palindromic* if it reads the same backwards as forwards, e.g., madam. In the next problem we are going to find all the palindromic words in a lot of languages! The approach is similar to Problem 4.13, however, in Problem 4.20 we will be working with strings of characters.

_____ Problem 4.20

Write a function to check whether a word is palindromic (symmetric). Using `DictionaryLookup`, find all the palindromic words in the English language. Then draw a bar chart showing how many palindromic words there are in the different languages which are supported by *Mathematica*.

$\Longrightarrow$ SOLUTION.

We first define a function to check whether a word is palindromic. Using
StringReverse, this is easy:

```
pal[n_] := n == StringReverse[n]
```

```
pal["test"]
False
```

```
pal["kayak"]
True
```

Now we select from the English dictionary available in *Mathematica* all the
words which are palindromic.

```
Select[DictionaryLookup[], pal]
{"a", "aha", "aka", "bib", "bob", "boob", "bub", "CFC",
"civic", "dad", "deed", "deified", "did", "dud", "DVD", "eke",
"ere", "eve", "ewe", "eye", "gag", "gig", "huh", "I", "kayak",
"kook", "level", "ma'am", "madam", "mam", "MGM", "minim",
"mom", "mum", "nan", "non", "noon", "nun", "oho", "pap",
"peep", "pep", "pip", "poop", "pop", "pup", "radar", "redder",
"refer", "repaper", "reviver", "rotor", "sagas", "sees", "seres",
"sexes", "shahs", "sis", "solos", "SOS", "stats", "stets", "tat",
"tenet", "TNT", "toot", "tot", "tut", "wow", "WWW"}
```

```
Length[Select[DictionaryLookup[], pal]]
70
```

This shows there are 70 palindromic words in this dictionary. We will do
the same with other languages, but first let us see what languages are available.

```
DictionaryLookup[All]
{"Arabic", "BrazilianPortuguese", "Breton",
"BritishEnglish", "Catalan", "Croatian", "Danish", "Dutch",
"English", "Esperanto", "Faroese", "Finnish", "French",
"Galician", "German", "Hebrew", "Hindi", "Hungarian",
"IrishGaelic", "Italian", "Latin", "Polish", "Portuguese",
"Russian", "ScottishGaelic", "Spanish", "Swedish"}
```

We will choose all the palindromic words from all these dictionaries. The
following is going to take about a minute.

```
t = Table[{lan, Length[Select[DictionaryLookup[{lan, All}],
pal]]},   {lan,DictionaryLookup[All]}]
```

```
{{"Arabic", 55}, {"BrazilianPortuguese", 81}, {"Breton",
   57}, {"BritishEnglish", 94}, {"Catalan", 150}, {"Croatian",
   106}, {"Danish", 157}, {"Dutch", 83}, {"English", 70},
   {"Esperanto", 14}, {"Faroese", 85}, {"Finnish", 72},
   {"French", 41}, {"Galician", 92}, {"German", 20},
   {"Hebrew", 159}, {"Hindi", 27}, {"Hungarian",
   146}, {"IrishGaelic", 42}, {"Italian", 22}, {"Latin",
   21}, {"Polish", 133}, {"Portuguese", 121}, {"Russian",
   25}, {"ScottishGaelic", 37}, {"Spanish", 58}, {"Swedish", 60}}
```

We are now ready to plot these on a Bar chart.

```
palnum = Last /@ t
{55, 81, 57, 94, 150, 106, 157, 83, 70, 14, 85, 72, 41, 92,
20, 159, 27, 146, 42, 22, 21, 133, 121, 25, 37, 58, 60}

BarChart[palnum, BarOrigin -> Left,
 ChartStyle -> "DarkRainbow", ChartLabels ->
 DictionaryLookup[All]]
```

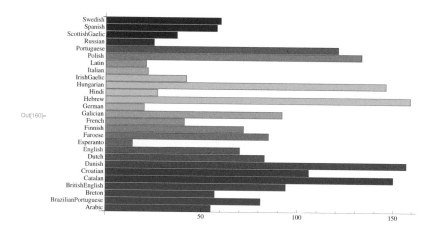

♣ TIPS

- In Problem 4.20 we used the command StringReverse, which reverses the order of the characters, so

```
StringReverse["this is great"]
"taerg si siht"
```

Mathematica has several more commands to work with strings, including StringLength, StringTake, StringDrop, StringReplace, ToString, etc.

——— Problem 4.21

Find all the words in the *Mathematica* dictionary such that the reverses of the words are in the dictionary as well (for example, loots and stool).

$\Longrightarrow$ SOLUTION.

There are 496 words in the *Mathematica* dictionary such that their reverses have a meaning as well. Here is the code:

```
Select[DictionaryLookup[], {StringReverse[#]} ==
    DictionaryLookup[StringReverse[#] ] &]
```

However, as we don't want to print all the words, we use Short to get one line of the list.

```
?Short
```

```
Short[expr] prints as a short form of expr, less than about
one line long.
Short[expr,n] prints as a form of expr about n lines long. >>
```

```
Short[
  Select[DictionaryLookup[], {StringReverse[#]} ==
    DictionaryLookup[StringReverse[#] ] &]]
```

```
{a,abut,agar,ah,aha,<<486>>,yaps,yard,yaw,yaws,yob}
```

└─────────────────────┘

We will give more interesting examples of this nature in Chapter 10, Pattern matching.

_____ Problem 4.22

The following code can be found among the examples for ExampleData in *Mathematica*. Explain what the code does.

```
Text[Row[
With[
{data = ExampleData[{"Text", "ToBeOrNotToBe"}, "Words"]},
 MapIndexed[Style[#, 2 Count[Take[data, First[#2]],#]] &, data]
 ],
  " "]]
```

$\Longrightarrow$ SOLUTION.

Here is the output of the code. It uses ExampleData to upload Hamlet's soliloquy with repeated words successively larger.

[decorative text block with scattered words: To, to, the, of, the, and, of, The, toh, The, of, the, The, ef, and, the, of, the, a, To, and, a, the, of, The, the, te, that we, of, us, And, the, of, the, of, And, of, and, And, the, of, Toh, in]

We analyse the code here. The code

```
data =
ExampleData[{"Text", "ToBeOrNotToBe"}, "Words"]
```

will load the text from Shakespeare's famous play, separate the words into a list and assign the result to data. The whole code is wrapped around by With which keeps data as the list of words throughout. We will look more closely at With and other commands to build blocks of codes in Section 11.1.

The code contains one pure function Style[#, 2 Count[Take[data, First[#2]],#]] &, which will be used by MapIndexed on the list data.

Since the text from Hamlet is long, let us work with another text, this one by Richard Feynman. We copy and paste this text into *Mathematica* and use built-in functions to make the words into a list.

```
x="all things are made of atoms, little particles that move
around in perpetual motion, attracting each other when they
are a little distance apart, but repelling upon being squeezed
into one another"

s = ReadList[StringToStream[x], Word]

{"all", "things", "are", "made", "of", "atoms,", "little",
"particles", "that", "move", "around", "in", "perpetual",
"motion,", "attracting", "each", "other", "when", "they",
"are", "a", "little", "distance", "apart,", "but", "repelling",
"upon", "being", "squeezed", "into", "one", "another"}
```

Next we explore the command MapIndexed.

```
MapIndexed[f, {a, b, c, d}]
{f[a, {1}], f[b, {2}], f[c, {3}], f[d, {4}]}

MapIndexed[First[#2] + f[#1] &, {a, b, c, d}]
{1 + f[a], 2 + f[b], 3 + f[c], 4 + f[d]}
```

As the example shows #2 gives the indices of each part. So we can use MapIndexed to get the location of each word in the sentence.

```
MapIndexed[{#, First[#2]} &, s]
```

```
{{"all", 1}, {"things", 2}, {"are", 3}, {"made", 4},
{"of", 5}, {"atoms,", 6}, {"little", 7}, {"particles", 8},
{"that", 9}, {"move", 10}, {"around", 11}, {"in", 12},
{"perpetual", 13}, {"motion,", 14}, {"attracting", 15},
{"each", 16}, {"other", 17}, {"when", 18}, {"they", 19},
{"are", 20}, {"a", 21}, {"little", 22}, {"distance", 23},
{"apart,", 24}, {"but", 25}, {"repelling", 26}, {"upon", 27},
{"being", 28}, {"squeezed", 29}, {"into", 30}, {"one", 31},
{"another", 32}}
```

Next, we will use Take[s,First[#2]] to get all the words up to a specific word and then count the number of times this word has been repeated up to where the word has appeared.

```
MapIndexed[Count[Take[s, First[#2]], #] &, s]
```

```
{1, 1, 1, 1, 1, 1, 1, 1, 1, 1, 1, 1, 1, 1, 1, 1, 1, 2,
1, 2, 1, 1, 1, 1, 1, 1, 1, 1, 1}
```

This list of numbers indicate the size of the font we will use, thus it will be passed to the command Style in Style[#, 2 Count[Take[data, First[#2]],#]] &. The rest of the code should be easily understood.

5

Changing heads!

Mathematica considers expressions in a unified manner. Any expression consists of a head and its arguments. For example {a,b,c} is considered as List[a,b,c] with the function List as the head and a,b,c as its arguments. *Mathematica* enables us to replace the head of an expression with another expression. For example, replacing the head of {a,b,c} with Plus gives Plus[a,b,c] which is a+b+c. This simple idea provides a powerful method to approach solving problems, as this chapter demonstrates.

Let us for a moment be a bit abstract. Wolfram *Mathematica*® has a very consistent way of dealing with expressions. Any expression in *Mathematica* has the following presentation head[arg1,arg2,...,argn] where head and arg could be expressions themselves. To make this point clear let us use the command FullForm which shows how *Mathematica* considers an expression.

```
FullForm[a + b + c]
Plus[a, b, c]

FullForm[a*b*c]
Times[a, b, c]

FullForm[{a, b, c}]
List[a, b, c]
```

Notice that the expressions a+b+c and {a,b,c} which represent very different things have such close presentations. Here Plus is a function and a,b,c are plugged into this function. Plus is the head of the expression a+b+c. One can see from the FullForm that the only difference between a+b+c and {a,b,c} is their heads! We can get the head of any expression:

```
Head[{a, b, c}]
List

Head[a + b + c]
Plus

{a, b, c}[[0]]
List
```

© Springer International Publishing Switzerland 2015
R. Hazrat, *Mathematica*®: *A Problem-Centered Approach*, Springer Undergraduate
Mathematics Series, DOI 10.1007/978-3-319-27585-7_5

Mathematica gives us the ability to replace the head of an expression with another head. The consequence of this is simply (head and) mind-blowing!

This can be done with the command `Apply`. Here is the traditional example:

```
Apply[Plus,{a,b,c}]
a+b+c
```

It is not difficult to explain this. The full form of {a,b,c} is List[a,b,c] with the head `List`. All *Mathematica* does is to change the head `List` to `Plus`, thus we have Plus[a,b,c] which is a+b+c.

The shorthand for `Apply` is `@@`, as the following example shows:

```
Plus @@ Range[10]
55
```

This gives the sum of 1 to 10.

Here are some more examples:

```
Times @@ Range[13] == 13!
True

Sin @@ Cos[x]
Sin[x]
```

To replace the head of the other layers of a list, one can specify the levels in the `Apply`. Study the following examples:

```
Apply[f, {{a, b}, {c}, {d, e}}]
f[{a, b}, {c}, {d, e}]

Apply[f, {{a, b}, {c}, {d, e}}, 1]
{f[a, b], f[c], f[d, e]}

f @@@ {{a, b}, {c}, {d, e}}
{f[a, b], f[c], f[d, e]}

Apply[f, {{a, b}, {c}, d}, 1]
{f[a, b], f[c], d}
```

───── Problem 5.1

Define the following functions:

$$cs(n) = 1^3 + 2^3 + \cdots + n^3$$
$$ep(n) = 1 + \frac{1}{1} + \frac{1}{2!} + \cdots + \frac{1}{n!}$$
$$p(n) = (1+x)(1+x^2)\cdots(1+x^n).$$

$\Longrightarrow$ SOLUTION.

Recall that `Range[n]` gives the list $\{1, 2, ..., n\}$ and since $\hat{}$ is a listable function, `Range[n]^3` produces $\{1^3, 2^3, ..., n^3\}$. Therefore, as in the example above, if we replace the head of the list with `Plus`, i.e., `Plus @@ Range[n]^3` we will have $1^3 + 2^3 + ... + n^3$. The other functions use a similar approach.

```
cs[n_]:=Plus @@ (Range[n]^3)
```

```
ep[n_]:=1.+Plus @@ (1/Range[n]!)
```

```
p[n_]:=Times @@ (1+x^Range[n])
```

Let us look at the second example. Again, `Range[n]` produces a list $\{1, 2, 3, \cdots, n\}$. Note that the factorial function `!` is listable, thus `Range[n]!` would produce $\{1!, 2!, 3!, \cdots, n!\}$. Recall that all the arithmetic operations are also listable, including `/`. Thus `1/Range[n]!` produces $\{\frac{1}{1!}, \frac{1}{2!}, \frac{1}{3!}, \cdots, \frac{1}{n!}\}$. We are almost there, all we have to do is to replace the head of $\{\frac{1}{1!}, \frac{1}{2!}, \frac{1}{3!}, \cdots, \frac{1}{n!}\}$ which is a `List` with `Plus` and as a result we get $\frac{1}{1!} + \frac{1}{2!} + \cdots + \frac{1}{n!}$.

All these functions are classical examples of using `Sum` and `Product` which are available in *Mathematica*. We will meet these commands in Chapter 7.

As the expressions get more complicated, it is quite difficult to analyze the `FullForm` of an expression (if this is necessary at all). The command `TreeForm` is an excellent facility to shed more light on how the functions are composed to get the expression. Here is an example:

```
FullForm[Sqrt[2 + x]/5 + x/y]
```

```
Plus[Times[Rational[1,5], Power[Plus[2,x], Rational[1,2]]],
Times[x, Power[y, -1]]]
```

```
TreeForm[Sqrt[2 + x] / 5 + x / y]
```

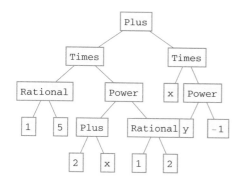

To understand this, start from the bottom of the tree and make your way to the top and you will get the expression

$$\frac{\sqrt{2+x}}{5} + \frac{x}{y}.$$

▬▬ Problem 5.2

Show that the only n less than 1000 such that

$$3^n + 4^n + \cdots + (n+2)^n = (n+3)^n$$

are the numbers 2 and 3.

⟹ SOLUTION.

We first need to write a code to evaluate $3^n + 4^n + \cdots + (n+2)^n$. This is very similar to the functions in Problem 5.1 and it looks like this

```
Apply[Plus, Range[3, n+2]^n]
```

Or, if using the shorthand, the equivalent code is `Plus @@ Range[3, n + 2]^ n`. We then select from the range of 1 to 1000, those n for which the result of the above is the same as `(n+3)^ n`. Note that in order to plug these into `Select` we need to combine these two into one anonymous function:

```
Select[Range[1000],Apply[Plus, Range[3, # + 2]^#] == (# + 3)^# &]
{2, 3}
```

┗▬▬▬▬▬▬▬

▬▬ Problem 5.3

Consider all numbers of the form $3n^2+n+1$, for positive integers $1 \le n \le 10000$. How small can the sum of the digits of such a number be? Can you find for which n this sum will be attained?

⟹ SOLUTION.

We start with a smaller range of 10 to 20. Once the code is sound, we apply it to the whole range of 1 to 10000. We use the pure function `3 #^2 + # + 1&` and send it into the range using `/@` to plug these numbers into the formula $3n^2 + n + 1$. After that we immediately use `IntegerDigits` to get the list of digits of each number. So our pure function takes the form `IntegerDigits[3 #^2 + # + 1]&`.

```
Range[10, 20]
{10, 11, 12, 13, 14, 15, 16, 17, 18, 19, 20}

IntegerDigits[3 #^2 + # + 1] & /@ Range[10, 20]
{{3, 1, 1}, {3, 7, 5},{4, 4, 5},{5, 2, 1}, {6, 0, 3}, {6, 9, 1},
{7, 8, 5}, {8, 8, 5}, {9, 9, 1}, {1, 1, 0, 3},  {1, 2, 2, 1}}
```

The next step is to sum the digits. This can be done by changing the head of each list by Plus. Thus we enhance our pure function to Plus @@ IntegerDigits[3 #^{2} + # + 1]&.

```
Plus @@  IntegerDigits[3 #^2 + # + 1] & /@ Range[10, 20]
{5, 15, 13, 8, 9, 16, 20, 21, 19, 5, 6}
```

Compare this with the code below. Can you explain the difference?

```
Plus @@@  (IntegerDigits[3 #^2 + # + 1] & /@ Range[10, 20])
{5, 15, 13, 8, 9, 16, 20, 21, 19, 5, 6}
```

Finally, we find the minimum value appearing in this list by using Min.

```
Min[Plus @@ IntegerDigits[3 #^2 + # + 1] & /@ Range[10000]]
3
```

To determine for which n this minimum is attained, we could write

```
t =Table[{i, Plus @@ IntegerDigits[3 i^2 + i + 1]}, {i, 1, 20}]

{{1, 5}, {2, 6}, {3, 4}, {4, 8}, {5, 9}, {6, 7}, {7, 11}, {8, 3},
{9, 10}, {10, 5}, {11, 15}, {12, 13}, {13, 8}, {14, 9}, {15, 16},
{16, 20}, {17, 21}, {18, 19}, {19, 5}, {20, 6}}

?SortBy
SortBy[list, f] sorts the elements of list in the order defined
by applying f to each of them

SortBy[t, Last]
{{8, 3}, {3, 4}, {1, 5}, {10, 5}, {19, 5}, {2, 6}, {20, 6},
{6, 7},  {4, 8}, {13, 8}, {5, 9}, {14, 9}, {9, 10}, {7, 11},
{12, 13}, {11, 15}, {15, 16}, {18, 19}, {16, 20}, {17, 21}}
```

Exercise 5.1

A number is called a *Harshad* number[1] if it is divisible by the sum of its digits (e.g., 12 is Harshad as it is divisible by 1+2=3). Find all 2-digit Harshad numbers. How many 5-digit Harshad numbers are there?

---

[1] Harshad means "giving joy" in Sanskrit, defined and named by the Indian mathematician D. Kaprekar.

Let's look at Problem 4.17 again.

——— Problem 5.4

Write a function squareFreeQ[n] that returns True if the number n is a square free number, and False otherwise.

⟹ SOLUTION.
    Here is the whole code:

```
squareFree1Q[n_] := Times @@ Last /@ FactorInteger[n] == 1

squareFree1Q /@ {12, 13, 14, 25, 26}
{False, True, True, False, True}
```

It might take a while to understand how this code works. If $n = p_1^{k_1} \cdots p_t^{k_t}$, then FactorInteger[n] will produce $\{\{p_1, k_1\}, \{p_2, k_2\}, \cdots, \{p_t, k_t\}\}$. We are after numbers such that all the k_i are 1 in the decomposition. Thus we can get all the k_i, multiply them together, and if we get anything other than 1, then this would be a non-square free number. Thus the first step is to apply Last to the list Last /@ FactorInteger[n] to get $\{k_1, k_2, \cdots, k_t\}$. Then all we have to do is to multiply them all together, and here comes the Times @@ to change the head of $\{k_1, k_2, \cdots, k_t\}$ from List to Times.

——— Problem 5.5

Find all the numbers up to one million which have the following property: if $n = d_1 d_2 \cdots d_k$ then $n = d_1! + d_2! + \cdots + d_k!$ (e.g. $145 = 1! + 4! + 5!$).

⟹ SOLUTION.

```
Select[Range[1000000], Plus @@ Factorial /@ IntegerDigits[#]
== # &]
{1, 2, 145, 40585}
```

The code consists of an anonymous function which, for any n, checks whether it has the desired property of the problem. Then, by using Select, we check the list of all the numbers from 1 to one million, Range[1000000]. Our anonymous function is Plus @@ Factorial /@ IntegerDigits[#] == # &. Let us look at the left-hand side of ==. The built-in function IntegerDigits[#] applied to $n = d_1 d_2 \cdots d_k$ produces the list of digits of n, namely $\{d_1, d_2, \cdots, d_k\}$. Next, applying Factorial /@ to this list, we get $\{d_1!, d_2!, \cdots, d_k!\}$. Now all we need is to find the sum of elements of this list, and this is possible by changing the head from `List` to `Plus` by `Plus @@`. Once this is done, we compare the left-hand side of `==` with the right-hand side, which is the original number `#`.

___

—— Problem 5.6

A number is *perfect* if it is equal to the sum of its proper divisors, e.g., $6 = 1 + 2 + 3$ but $18 \neq 1 + 2 + 3 + 6 + 9$. Write a program to find all the perfect numbers up to 10000. (Hint, have a look at the command `Divisors`.)

$\Longrightarrow$ SOLUTION.

Here is a step-by-step approach to the solution.

```
Divisors[6]
{1,2,3,6}

Most[Divisors[6]]
{1,2,3}

Apply[Plus,Most[Divisors[6]]]
6

Select[Range[10000], # == Apply[Plus, Most[Divisors[#]]] &]
{6, 28, 496, 8128}
```

The numbers 6, 28 and 496 were already known as being perfect numbers 2000 years before Christ. A glance at the list shows that all the perfect numbers we have found are even. It is still unknown whether there are any odd perfect numbers. This is probably the oldest unsolved question in mathematics!

___

—— Problem 5.7

Among the first 100000 numbers, what is the largest number n which is divisible by all positive integers $\leq \sqrt{n}$?

$\Longrightarrow$ SOLUTION.

```
Select[Range[100000],(Mod[#,LCM @@ Range[Floor[Sqrt[#]]]]==0)&]
{1, 2, 3, 4, 6, 8, 12, 24}
```

___

Exercise 5.2

Decipher what the following codes do:

```
g[n_] := Times @@ Apply[Plus , Inner[List, x^Range[n],
1/x^Range[n], List], 1]

t[n_] := Times @@ Apply[Plus, Thread[List[x^Range[n],
1/x^Range[n]]], 1]
```

Exercise 5.3

Find all the numbers up to 100 which have the following property: if $n = p_1^{k_1} \cdots p_t^{k_t}$ is the prime decomposition of n then $n = k_1 \times p_1 + k_2 \times p_2 + \cdots + k_t \times p_t$.

Exercise 5.4

Decipher what the following codes do:

```
Power @@ (x + y)

Plus @@@ (x^y + y^z)

Times @@ Power @@@ FactorInteger[45]
```

Exercise 5.5

Show that the sum of all the divisors of the number

$$6086555670238378989670371734243169622657830773351885970528324860512791691264$$

is a perfect number. (This number is the only known sublime number besides 12. A *sublime number* is a positive integer which has a perfect number of positive divisors (including itself), and whose positive divisors add up to another perfect number. See Problem 5.6 for perfect numbers and MathWorld [9] for sublime numbers.)

Exercise 5.6

Find all the positive integers n less than 20 such that the Euclid's formula $2^{n-1}(2^n - 1)$ produces perfect numbers.

Exercise 5.7

A *weird number* is a number such that the sum of its proper divisors (divisors including 1 but not itself) is greater than the number, but no subset of those divisors sums to the number itself. Find all the weird numbers up to 100.

6

A bit of logic and set theory

This chapter focuses on statements which take the values of True or False, i.e., Boolean statements. We can combine statements of this type to make decisions based on their values. We introduce the If statement available in *Mathematica*.

6.1 Being logical

In mathematical logic, statements can have a value of True, False or undefined.[1] These are called *Boolean expressions*. This helps us to "make a decision" and write programs based on the value of a statement (I am thinking of the classical If-Then statement; If something is True, Then do this, Otherwise do that). We have seen ==, which compares the left-hand side with the right-hand side. Studying the following examples carefully will tell us how Wolfram *Mathematica*® approaches logical statements:

```
3^2+4^2==5^2
True

3^2+4^2>5^2
False

9Sqrt[10!] < 10Sqrt[9!]
False

(x-1)(x+1)==x^2-1
(x-1)(x+1)==x^2-1
```

---

[1] We don't want to go into detail here, mainly because I don't know the detail!

© Springer International Publishing Switzerland 2015
R. Hazrat, *Mathematica®: A Problem-Centered Approach*, Springer Undergraduate
Mathematics Series, DOI 10.1007/978-3-319-27585-7_6

```
Simplify[%]
True

x==5
x==5

{1,2}=={2,1}
False

{a,b}=={b,a}
{a,b}=={b,a}
```

As one notices, *Mathematica* echoes back the expressions that she can't evaluate (e.g., x==5). Among them is {a,b}=={b,a}, although one expects to get False as lists respect order. This is because *Mathematica* does not know about the values of a and b, and in case a and b are the same then {a,b}=={b,a} is True, and False otherwise. If you want *Mathematica* to judge from the face value, then use ===,

```
x==5
False

{a,b}==={b,a}
False

?===
ihs===rhs yields True if the expression lhs is identical
to rhs and yields False otherwise. >>
```

One can combine logical statements with the usual boolean operations And, Or, Not, or the equivalent &&, ||, ! as the following examples show:

```
2 > 3 && 3 > 2
False

And[2 > 3, 3 > 2]
False

1 < 2 < 3
True

2 > 3 || 2 < 3
True

Or[2 > 3, 3 > 2]
True

Not[Not[2>1]]
True

3^2+4^2>= 5^2
True
```

In general, for two statements $\mathcal{A}$ and $\mathcal{B}$, the statement $\mathcal{A}$&&$\mathcal{B}$ is false if one of $\mathcal{A}$ or $\mathcal{B}$ is false and $\mathcal{A} \parallel \mathcal{B}$ is true if one of them is true. In order to produce all possible combinations of true and false cases, we use the command Outer as the following example shows (we will look at the command Outer in more detail in Section 8.5).

```
Outer[f, {a, b}, {x, y}]
{{f[a, x], f[a, y]}, {f[b, x], f[b, y]}}
```

Thus, if in the above we replace f by And or Or we will get all the possible combinations of True and False.

```
Outer[And, {True, False}, {True, False}]
{{True, False},{False,False}}
```

```
Outer[Or, {True, False}, {True, False}]
{{True, True}, {True,False}}
```

In *Mathematica*, for a variable, one can specify certain domains. This means that the variable takes its values from a specific type of data. The domains available are Algebraics, Booleans, Complexes, Integers, Primes, Rationals and Reals. One of the fundamental theorems in number theory is to show that π is not a rational number, i.e., is not of the form m/n, where m and n are integers. Look at the following examples:

```
Pi ∈ Rationals
False
```

```
Sqrt[7] ∈ Integers
False
```

—— Problem 6.1

Show that $1 + \sqrt{3} + \sqrt{5} + \sqrt{7}$ is an algebraic number (i.e., it is a solution of a polynomial equation with integer coefficients).

$\Longrightarrow$ SOLUTION.

```
Plus @@ Sqrt[Range[1, 7, 2]] ∈ Algebraics
True
```

One can use membership ($\in$) to write neat solutions to several problems.

—— Problem 6.2

Does the formula $(n!)^2 + 1$ give prime numbers for $n = 1$ to 6?

$\Longrightarrow$ SOLUTION.

```
(#!^2 + 1) & /@ Range[6] ∈ Primes
False
```

Here we first apply the anonymous function (#!^2 + 1), which is the formula $(n!)^2 + 1$, to the list containing 1 to 6. Then we ask *Mathematica* if the elements of this list belong to the domain Primes. The answer is False. The following code shows that the above formula does not produce a prime number for $n = 6$:

```
PrimeQ /@ ((#!^2 + 1) & /@ Range[1, 6])
{True, True, True, True, True, False}
```

One should know that *Mathematica* cannot (yet) perform miracles. For example, one can actually prove that $\sqrt[3]{2 + \sqrt{5}} + \sqrt[3]{2 - \sqrt{5}}$ is an integer, but *Mathematica* tells us

```
(2 + 5^(1/2))^(1/3)+(2 - 5^(1/2))^(1/3) ∈ Integers
False
```

```
FullSimplify[(2 + 5^(1/2))^(1/3) +
        (2 - 5^(1/2))^(1/3) ∈ Integers]
False
```

Mathematica provides the logical quantifiers ∀, ∃ and ⇒ with ForAll, Exist and Implies commands. But these seem not to be that powerful yet. For example, using these commands one cannot prove Fermat's little theorem, which says $2^{p-1} \equiv 1 \pmod p$ where $p > 2$ is a prime number!

```
ForAll[p, p ∈ Primes, Mod[2^(p - 1), p] == 1]
```

```
FullSimplify[ForAll[p, p ∈ Primes, Mod[2^(p - 1),
p] == 1]]
```

Nor can we use them to prove the easy fact that the product of four consecutive numbers plus one is a square number.

```
Implies[n ∈ Integers && n > 0, Sqrt[n(n + 1)(n +
2)(n + 3) + 1] ∈ Integers]
```

In both cases *Mathematica* gives back the same expression, indicating she cannot decide on them.

6.2 Handling sets

Now it has been agreed among mathematicians that *any* mathematics starts by considering sets, i.e., collections of objects.[2] As we mentioned, the difference between mathematical sets and lists in *Mathematica* is that lists respect order and repetition, which is to say one can have several copies of one object in a list (see Chapter 4). Sets are not sensitive about repeated objects, e.g., the set $\{a, b\}$ is the same as the set $\{a, b, b, a\}$. There is no concept of sets in *Mathematica* and if necessary one considers a list as a set.

If one wants to get rid of duplications in a list, one can use

```
Union[{a,b,b,a}]
{a,b}
```

Considering two sets, the natural operations between them are union and intersection. *Mathematica* provides Union to collect all elements from different lists into one list (after removing all the duplications) and Intersection for collecting common elements (again discarding repeated elements). The following examples show how these commands work.

```
u= {1, 2, 3, 4, 5, 2, 4, 7, 4}; a = {1, 4, 7, 3};
   b = {5, 4, 3, 2};

Union[u]
{1, 2, 3, 4, 5, 7}

a ∪ b

{1, 2, 3, 4, 5, 7}

a ∩ b

{3,4}

Complement[u, a]
{2, 5}

?Complement
Complement[eall, e1, e2, ... ] gives the elements in
eall which are not in any of the ei.
```

---

[2] To be precise, one first considers classes.

```
Complement[u, a ∩ b] == Complement[u, a] ∪
Complement[u, b]
True
```

The first example shows Union[list] will get rid of repetition in a list. The command Complement[u,a] will give the elements of u which are not in a. From the example one can see that a ∩ b is acceptable in *Mathematica* and is a shorthand for Intersection[a,b]. In the last example we checked a theorem of set theory that $(A \cap B)^c = A^c \cup B^c$, where ^{c} stands for complement.

Example 6.3

The following trick will be used later (in Chapter 8, inside a loop) to collect data.

```
A={}
A=A ∪ {x}
{x}
A=A ∪ {y}
{x,y}
A=A ∪ {z}
{x,y,z}
```

This is similar to the traditional trick sum=sum+i. Each time sum=sum+i is performed, i will be added to sum and this result will be the new value of sum.

There are other ways to add an element to a list.

```
Append[{a,b,c},d]
{a,b,c,d}

A={};A=Append[A,x]
{x}

A=Append[A,y]
{x,y}

A=Append[A,z]
{x,y,z}
```

Also note that the command AppendTo[s, elem] is equivalent to s = Append[s, elem].

▬▬ Problem 6.4

How many positive integers n are there such that n is a divisor of at least one of the numbers 10^{40} and 20^{30}?

$\Longrightarrow$ SOLUTION.

Recall that `Divisors[n]` will produce a list of all the positive integers which divide n. So all we need to do is to get the collection of divisors of 10^{40} and 20^{30}. This can be done with `Union`. Note that we are not going to print out all the divisors as this is a long list.

```
Length[Union[Divisors[10^40], Divisors[20^30]]]
2301
```

—— Problem 6.5

Find out how many common words there are between the English language and French, Dutch and German respectively.

$\Longrightarrow$ SOLUTION.

Recall that `DictionaryLookup` contains all the words in 26 languages (see Problem 4.20). All we have to do is to intersect the list of words in the English language with the list of French words and so on as follows:

```
Length[Intersection[DictionaryLookup[{"English", All}],
  DictionaryLookup[{"French", All}]]]
6897

Length[Intersection[DictionaryLookup[{"English", All}],
  DictionaryLookup[{"Dutch", All}]]]
6166

Length[
 Intersection[DictionaryLookup[{"English", All}],
  DictionaryLookup[{"German", All}]]]
1286
```

♣ TIPS

– The command `Tally[list]` tallies the elements in `list`, listing all distinct elements together with their multiplicities.

– The command `Join[list1,list2]` concatenates lists or other expressions that share the same head.

6.3 Decision making, If and Which

The statement If[stat,this,that] where stat is a Boolean expression, i.e., has the value of True or False, will execute this if the stat value is True and that otherwise. That means, in either case, one of the statements this or that will be performed (but not both). So this gives us the ability to make a decision about which part one wants to perform. Here is an example:

```
If[12^13+13>13^12+12,
Print["12^13+13>13^12+12"],
Print["12^13+13<13^12+12"]
]
12^13+13>13^12+12
```

This shows that $12^{13} + 13 > 13^{12} + 12$. However, the reader should be cautious, since if it happened that $12^{13} + 13 = 13^{12} + 12$, the output would still have been $12^{13} + 13 < 13^{12} + 12$ (why?). In such situations where there might be more possibilities, the command Which is better suited. Study this example

```
Which[
  12^13 + 13 > 13^12 + 12, Print["12^13+13>13^12+12"],
  12^13 + 13 < 13^12 + 12, Print["12^13+13<13^12+12"],
  12^13 + 13 == 13^12 + 12, Print["12^13+13=13^12+12"]
]
12^13+13>13^12+12
```

Using If or Which, we are now able to define functions which have conditions.

──── **Problem 6.6**

Define the Collatz function as follows:

$$f(x) = \begin{cases} x/2 & \text{if } x \text{ is even} \\ 3x + 1 & \text{if } x \text{ is odd.} \end{cases}$$

It was conjectured by L. Collatz in 1937 that if one applies f repeatedly to any positive integer, one eventually arrives at 1. Find out how many times one needs to apply f to 16 in order to reach 1.

$\Longrightarrow$ SOLUTION.

Recall that EvenQ is a Boolean statement which returns True if the number is even and False otherwise. The function f consists of two parts: If n is even, then $f(n) = x/2$; otherwise $f(n) = 3n + 1$. One can use an If statement to define this function as follows:

```
f[n_] := If[EvenQ[n], n/2, 3 n + 1]

f[16]
8

f[8]
4

f[4]
2

f[2]
1
```

We will return to this conjecture and will write this function again in Problems 10.9 and 11.2 using the *Mathematica* rules and functions with multiple definitions, respectively. For a comprehensive discussion of the Collatz function using *Mathematica* see also Chapter 7 of Vardi's book [6].

▌▬▬▬▬▬▬▬

There are situations in which one needs to look at several possibilities (so Which would be a good tool) and if none of the possibilities occur, then as a last resort, one carries on with a specific case. This will be demonstrated in the next problem.

━━━ Problem 6.7

Define the function

$$f(x) = \begin{cases} -x, & \text{if } |x| < 1 \\ \sin(x), & \text{if } 1 \le |x| < 2 \\ \cos(x), & \text{otherwise.} \end{cases} \tag{6.1}$$

⟹ SOLUTION.
  The function $f(x)$ is defined as follows: if $|x| < 1$ or $1 \le |x| < 2$ then $f(x) = -x$ or $f(x) = \sin(x)$, respectively. However, if x does not fall into any of the cases above, then $f(x)$ is defined as $\cos(x)$. Here is how we handle this using Which:

```
f[x_] := Which[
  Abs[x] < 1, -x,
  1 <= Abs[x] < 2, Sin[x],
  True, Cos[x]
  ]
```

As you may guess, $|x|$ is translated into *Mathematica* using the Abs function. Here is a little test for the function f:

```
f /@ {0.5, Pi/2, Pi}
{-0.5, 1, -1}
```

There is also another way to define this function using the command Piecewise.

```
g[x_] := Piecewise[{{-x, Abs[x] < 1}, {Sin[x], 1 <= Abs[x] < 2}},
   Cos[x]]

g /@ {0.5, Pi/2, Pi}
{-0.5, 1, -1}
```

There is an important difference between these two definitions which will be explored in Problem 14.6. Basically, defining the function using Which makes *Mathematica* consider this function as a continuous function. We can rectify this problem using Piecewise.

▌▬▬▬▬▬▬▬▬▬

▬▬▬ Problem 6.8

Define the function

$$f(x,y) = \begin{cases} e^x & \text{if } x = y \\ \frac{e^x - e^y}{x-y} & \text{if } x \neq y \end{cases}$$

and observe that for arbitrary real numbers a, b such that $a < b < 0$ we have

$$\frac{f(x,b)}{f(x,a)} > \frac{1+e^b}{1+e^a}$$

for any $a \leq x \leq b$.

⟹ SOLUTION.

Here we have a function with more than one definition: if $x = y$ then $f(x,y) = e^x$; otherwise $f(x,y) = \frac{e^x - e^y}{x-y}$. Thus the definition of this function calls for the If statement.

```
ec[x_, y_] := If[x == y, E^x, (E^x - E^y)/(x - y)]
```

We will use Plot to check the claim of the problem. For graphics, see Chapter 14.

```
a = -9; b = -5;
```

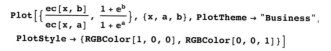

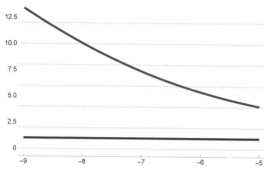

Figure 6.1 Graphs of $\frac{f(x,10)}{f(x,9)}$ and $\frac{1+e^{-5}}{1+e^{-9}}$

7

Sums and products

This chapter is devoted to evaluating series. *Mathematica* provides two commands, Sum and Product (and their numerical cousins, NSum, NProduct) to handle series.

In the previous chapters we could write a code to calculate the series

$$1 + \frac{1}{1} + \frac{1}{2!} + \cdots + \frac{1}{n!}.$$

Wolfram *Mathematica*® offers us two commands, namely Sum and Product, to handle problems of this nature easily without getting into any programming complications.

7.1 Sum

Study the following examples:

```
Sum[s[i], {i, 1, 7}]
s[1] + s[2] + s[3] + s[4] + s[5] + s[6] + s[7]

Sum[s[i], {i, 1, k}]
```
$\sum_{i=1}^{k} s[i]$

The second example shows again that *Mathematica* can handle things symbolically.

© Springer International Publishing Switzerland 2015
R. Hazrat, *Mathematica*®: *A Problem-Centered Approach*, Springer Undergraduate
Mathematics Series, DOI 10.1007/978-3-319-27585-7_7

———— Problem 7.1

Write the function $ep(n) = 1 + \dfrac{1}{1} + \dfrac{1}{2!} + \cdots + \dfrac{1}{n!}$.

$\Longrightarrow$ SOLUTION.

Here is the code:

```
ep[n_] := 1 + Sum[1/k!, {k, 1, n}]

N[ep[100]]
2.71828

N[E]
2.71828
```

Sometimes *Mathematica* can do great things:

```
ep[Infinity]
E
```

This shows that the above series converges to $\mathbb{E}$, i.e.,

$$1 + \sum_{i=1}^{\infty} \frac{1}{n!} = \mathbb{E},$$

and *Mathematica* is aware of this.

———— Problem 7.2

Prove that

$$(1 + 2 + 3 + \cdots + n)^2 = (1^3 + 2^3 + 3^3 + \cdots + n^3).$$

$\Longrightarrow$ SOLUTION.

Writing the above equality symbolically, we want to show $\sum_{i=1}^{n} i^3 = (\sum_{i=1}^{n} i)^2$. This example shows that *Mathematica* is aware of formulas for certain sums, including the ones above:

```
p[n_] := Sum[i, {i, 1, n}]

p[n]
(1/2) n (1 + n)

p3[n_] := Sum[i^3, {i, 1, n}]

p3[n]
(1/4) n^2 (1+n)^2

p[n]^2==p3[n]
True
```

The above example shows *Mathematica* knows that $1+2+\cdots+n = \frac{n(n+1)}{2}$ and $1^3 + 2^3 + \cdots + n^3 = (\frac{n(n+1)}{2})^2$. The first formula was known to Gauss at the age of seven. In fact he proved the formula as follows:

$$
\begin{array}{ccccc}
1 & 2 & \cdots & n & + \\
n & n-1 & \cdots & 1 & \\
\hline
n+1 & n+1 & \cdots & n+1 &
\end{array}
$$

Thus twice the sum of the series is $n(n+1)$ and thus the formula. The second formula follows by an easy induction.

Exercise 7.1

Evaluate $\sum_{k=1}^{n} \frac{k}{k^4+k^2+1}$.

Problem 7.3

Write a function to calculate the following series

$$s(n) = \frac{1}{1} + \frac{1}{1+2} + \dots + \frac{1}{1+2+\dots+n}$$

$\Longrightarrow$ SOLUTION.

A glance at the series shows that there are in fact two series involved. Thus one needs to use Sum twice, once to take care of $1 + 2 + \cdots + i$ and again for the sum of these expressions.

```
s[n_] := Sum[1/Sum[j, {j, 1, i}], {i, 1, n}]
```

One probably needs a few minutes to be convinced that this code generates the series given in the problem. One of the advantages of the Front-End in *Mathematica* is its ability to write mathematics as one writes it on paper. Writing the above series using mathematical symbols, one has $\sum_{i=1}^{n}(1/\sum_{j=1}^{i}j)$, which is much more understandable than $s[n]$.

Using the palette provided by *Mathematica*, one can enter exactly the same expression in the front-end and define the function s this way.

$$s[n\_] = \sum_{i=1}^{n}(1/\sum_{j=1}^{i}j)$$

Mathematica can easily handle complicated symbolic calculations, as the following example demonstrates. Recall that the binomial coefficient $\binom{n}{k}$ stands for $\frac{n!}{k!(n-k)!}$. The command Binomial[n, k] is available (see Problem 2.6).

──── Problem 7.4

Define

$$p(n) = \sum_{k=0}^{n}\binom{n}{k}^{2}(1+x)^{2n-2k}(1-x)^{2k}$$

and show that, for any chosen n, the coefficients of x are non-negative.

⟹ SOLUTION.

We shall first carefully translate the above formula into *Mathematica*.

```
p[n_] := Sum[Binomial[n, k]^2(1 + x)^(2n - 2k)(1 - x)^(2k),
{k, 0, n}]

p[3]
(1 - x)^6 + 9(1 - x)^4(1 + x)^2 + 9(1 - x)^2(1 + x)^4 + (1+x)^6

Expand[p[3]]
20 + 12x^2 + 12x^4 + 20x^6
```

As one sees, all the coefficients are non-negative. One can gather these coefficients in a list

```
CoefficientList[Expand[p[7]], x]

{3432, 0, 1848, 0, 1512, 0, 1400, 0, 1400, 0, 1512, 0, 1848, 0,
3432}
```

This is one of my favourite examples of using Sum.

------ **Problem 7.5**

Define $S(k,n) = \sum_{i=1}^{n} i^k$. Prove that

$$\sum_{a=0}^{n} \frac{S(2, 3a+1)}{S(1, 3a+1)} \qquad (7.1)$$

is always a square number.

⟹ SOLUTION.

The function s takes two variables and is defined as a series. Once we have defined this function, we can simply plug it into Equation 7.1.

```
s[k_, n_] := Sum[i^k, {i, 1, n}]

Sum[s[2, 3a + 1]/s[1, 3a +1],{a,0, n}]

1 + n + n (1 + n)

Factor[%]
(1 + n)^2
```

♣ TIPS

− Sum will try to evaluate the precise sum of the series. If an approximation suffices, use NSum, which is often much faster.

We finish this section by observing "live" that the series

$$\frac{1}{1} + \frac{1}{4} + \frac{1}{9} + \frac{1}{25} + \cdots$$

conversges to $\frac{\pi^2}{6}$.

```
Manipulate[Pi^2 / 6 - NSum[1 / i^2, {i, 1, n}], {n, 1, 100 000, 1}]
```

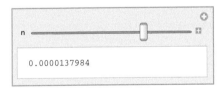

```
0.0000137984
```

Exercise 7.2

Define the functions $f(k) = \sum_{n=1}^{k}(-1)^n \frac{t^n}{n!}$ and $g(k) = \sum_{n=1}^{k}(-1)^n \frac{n!}{t^n}$.
Show that

$$2 - f(2)g(2) = \frac{2}{t} + \frac{t}{2}.$$

Exercise 7.3

Write the function

$$f(k) = \sin(x) + x + \frac{x^3}{1 \times 2 \times 3} + \frac{x^5}{1 \times 2 \times 3 \times 4 \times 5} + \cdots + \frac{x^k}{1 \times 2 \times \cdots \times k}$$

(k is odd).

Exercise 7.4

Verify that

$$\sum_{n=0}^{\infty} \frac{(-1)^n}{2n+1} \sum_{k=0}^{2n} \frac{1}{2n + 4k + 3} = \frac{3\pi}{8} \log \frac{\sqrt{5}+1}{2} - \frac{\pi}{16} \log 5$$

Exercise 7.5

Investigate whether the series

$$1 + \frac{1}{2} + \frac{1}{3} - \frac{1}{4} - \frac{1}{5} - \frac{1}{6} + \frac{1}{7} + \frac{1}{8} + \frac{1}{9} - - - + + + \cdots$$

converges.

Exercise 7.6

Investigate whether the series

$$2^{-\frac{1}{2}} + (3+5)^{-\frac{1}{2}} + (7+11+13)^{-\frac{1}{2}} + (17+19+23+29)^{-\frac{1}{2}} + \cdots$$

converges.

7.2 Product

The command Product is very similar to Sum, but here instead of a sum we have a series involving products:

```
Product[s[i], {i, 1, 7}]
s[1]s[2]s[3]s[4]s[5]s[6]s[7]
```

```
Product[s[i], {i, 1, k}]
```
$\prod_{i=1}^{k} s[i]$

Here is a code to produce $(x + \frac{1}{x})(x^2 + \frac{1}{x^2}) \cdots (x^n + \frac{1}{x^n})$.

```
p[n_] := Product[(x^k + 1/x^k), {k, 1, n}]
```

_____ Problem 7.6

Write the following series:

$$\frac{1 \times 3}{2 \times 2}\frac{3 \times 5}{4 \times 4}\frac{5 \times 7}{6 \times 6} \cdots$$

and show that this series tends to $2/\pi$.

$\Longrightarrow$ SOLUTION.

First one needs to recognise that the general term of this series is

$$\frac{(2n-1)(2n+1)}{2n \times 2n},$$

i.e.,

$$\prod_{n=1} \frac{(2n-1)(2n+1)}{2n \times 2n} = \frac{1 \times 3}{2 \times 2}\frac{3 \times 5}{4 \times 4}\frac{5 \times 7}{6 \times 6} \cdots.$$

Knowing this, it is easy to write down the code

```
N[Product[(2 n - 1) (2 n + 1)/(2 n)^2, {n, 1, 100}] - 2/Pi]
0.0015856

N[Product[(2 n - 1) (2 n + 1)/(2 n)^2, {n, 1, 1000}] - 2/Pi]
0.000159095

N[Product[(2 n - 1) (2 n + 1)/(2 n)^2, {n, 1, 10000}] - 2/Pi]
0.0000159149
```

This shows that as n grows, the product gets closer to $\frac{2}{\pi}$. The punch line is:

```
Product[(2 n - 1) (2 n + 1)/(2 n)^2, {n, 1, Infinity}]
2/Pi
```

└─────────────────

♣ TIPS

– Product will try to evaluate the precise product of the series. If an approximation suffices, use NProduct, which is often much faster.

Exercise 7.7

Let F_i be the i-th Fibonacci number. Write the function

$$f(n) = (F_1 + x)(F_2 + x^2) \cdots (F_n + x^n).$$

What is the coefficient of x^4 in $f(23)$? (Hint: see Coefficient.)

Exercise 7.8

Investigate that

$$\frac{\exp}{2} = \left(\frac{2}{1}\right)^{\frac{1}{2}} \left(\frac{2}{3}\frac{4}{3}\right)^{\frac{1}{4}} \left(\frac{4}{5}\frac{6}{5}\frac{6}{7}\frac{8}{7}\right)^{\frac{1}{8}} \left(\frac{8}{9}\frac{10}{9}\frac{10}{11}\frac{12}{11}\frac{12}{13}\frac{14}{13}\frac{14}{15}\frac{16}{15}\right)^{\frac{1}{16}} \dots$$

8
Loops and repetitions

In this chapter we will look at traditional loops available in *Mathematica*, i.e., ways to repeat a block of code a number of times. We will introduce Do, While and For loops and study nested loops, that is, loops defined inside each other. We finish the chapter by looking at other nested commands.

If we agree that the primary ability that a computer language provides is the ability to repeat a certain code "fast" then Wolfram *Mathematica*® provides three *loops* that enable us to repeat part of our codes. These are quite similar to the loops that exist in any procedural language like Pascal or C.

8.1 Do, For a While

The first and the simplest one is the Do loop. Here is the traditional example. The structure of the Do loop reminds one of the commands such as Sum or Table.

```
Do[Print[i],{i,1,7}]
1
2
3
4
5
6
7
```

The code:

```
Do[f[i],
{i,1,1000000}]
```

© Springer International Publishing Switzerland 2015
R. Hazrat, *Mathematica®: A Problem-Centered Approach*, Springer Undergraduate Mathematics Series, DOI 10.1007/978-3-319-27585-7_8

repeats the expression f[i] one million times where i runs from 1 to 1000000. In fact this is (almost) equivalent to

```
f/@ Range[10000000];
```

Here is a little comparison.

```
Timing[Do[f[i],{i,1,10000000}]]
{6.93 Second,Null}
```

```
Timing[f/@ Range[10000000];]
{5.008 Second,Null}
```

Apart from writing a code which is faster, the art of programming is also to try to write codes in a way in which they are also readable.

We will write Problem 4.19 using a Do loop.

━━━ Problem 8.1

Find out how many primes bigger than n and smaller than $2n$ exist, when n runs from 1 to 15.

$\Longrightarrow$ SOLUTION.

First we find all prime numbers up to 30.

```
prime30=Select[Range[30],PrimeQ]
{2,3,5,7,11,13,17,19,23,29}
```

Now for any n we check how many prime numbers lie between n and $2n$. To do this, as in Problem 4.19, we create a list of numbers between n and $2n$, Range[n+1,2n-1], then using Intersection we find out how many prime numbers are in this interval, Range[n+1,2n-1] ∩ prime60. Using Length we can find the number of primes that lie in this interval. Once we have this, then using a Do loop, we run this code for n from 1 to 15.

```
Do[
Print[n, " ::::", Length[Intersection[Range[n + 1, 2 n - 1],
prime60]]],
{n, 1, 15}]
1::::0
2::::1
3::::1
4::::2

5::::1
6::::2
7::::2
8::::2
```

```
 9::::3
10::::4
11::::3
12::::4
13::::3
14::::3
15::::4
```

For our next application of a Do loop, recall Problem 4.8. The formula $n^2 + n + 41$ produces prime numbers when n runs from 0 to 39. This was noticed by Euler some 300 years ago. One wonders whether one gets more consecutive prime numbers for a different constant in the above formula. The next problem examines this:

——— Problem 8.2

Consider the formula $n^2 + n + i$. Find out the number of consecutive primes (starting from $n = 0$) that one gets when i runs from 1 to 10000.

$\Longrightarrow$ SOLUTION.

One way to approach the problem is to write a code to find out how many consecutive primes one gets (starting from $n = 0$) for a fixed i in the formula $x^2 + x + i$. Once this is done, then one can use a Do loop to change the value of i from 1 up to 10000. The code which finds out the number of consecutive primes is similar in nature to that given in Problem 4.17.

The code

```
Select[Range[500], !PrimeQ[#^2 + # + 41] &, 1]
{40}
```

returns the first number in the range of $\{0, \cdots, 500\}$ such that the formula $n^2 + n + 41$ does not return a prime number. All we have to do now is to assemble this code in a loop as follows:

```
A={};
Do[
A=Union[A,Select[Range[500], !PrimeQ[#^2+#+i]&,1]],
{i,1,10000}];A

{1,2,3,4,5,6,7,10,12,16,40}
```

A line such as

```
A=Union[A,Select[Range[100], !PrimeQ[#^2+#+i]&,1]]
```

which is equivalent to

```
A= A ⋃ Select[Range[100],!PrimeQ[#^2+#+i]&,1]
```

collects all new results in the list `A`. We have seen this trick in Example 6.3.

A glance at the result shows that n^2+n+41 produces the maximum number of consecutive primes, as was noticed by Euler. As a matter of fact, the formula which produces 16 consecutive prime numbers is $n^2 + n + 17$, which was also found by Euler!

∟‾‾‾‾‾‾‾‾‾

Exercise 8.1

Modify the code of Problem 8.2 to find for which values of i one gets $\{10, 12, 16, 40\}$ consecutive prime numbers from the formula $n^2 + n + i$.

——— Problem 8.3

The sum of two positive integers is 5432 and their least common multiple is 223020. Find the numbers.

⟹ SOLUTION.

```
Do[
 If[LCM[i, 5432 - i] == 223020, Print[i, "  ", 5432-i]],
 {i, 1, 2718}]

1652   3780
```

∟‾‾‾‾‾‾‾‾‾

The following `Do` loop will calculate a continued fraction. We will visit these fractions again using recursive functions in Problem 12.6.

——— Problem 8.4

Given non-negative integers $c_1, c_2, \ldots, c_m$, where $c_i \geq 0$ for $i \geq 2$, define

$$[c_1, c_2, \ldots, c_m] = c_1 + \cfrac{1}{c_2 + \cfrac{1}{\cdots + \cfrac{1}{c_m}}}.$$

Using a Do loop, write a function to get a list $\{c_1, c_2, \ldots, c_m\}$ and calculate the continued fraction $[c_1, c_2, \ldots, c_m]$.

$\Longrightarrow$ SOLUTION.

We create a Do loop which adds the fractions starting from $1/c_m$ and then adds the next member of the list on the left, i.e., c_{m-1} to $1/c_m$, and then repeating this process $m - 1$ times.

```
g[s_List] :=
  {
    c = s[[-1]];
    Do[c = s[[i]] + 1/c,
    {i, Length[s] - 1, 1, -1}]; c
    }
```

```
g[{1, 2, 3}]
{10/7}
```

Note that the function consists of three lines, separated from each other by a semicolon ;. For this reason, {} wraps around the whole block. This is also the reason the answer is shown as $\{10/7\}$. Later, in Section 11.1, we will see the command Module is a better choice when developing a block of codes.

We shall see more examples of the Do loop later. The next loop is the While loop. This one operates on a Boolean (True or False) statement and gives you the ability to repeat a block until the Boolean statement becomes False.

The While loop has the form

$$\text{While[condition,body].}$$

The body of the loop can consist of several lines separated by ;.

▬▬ Problem 8.5

Find the first prime number consisting only of ones and greater than 11.

$\Longrightarrow$ SOLUTION.

Here is the mystery code:

```
n=111;
While[!PrimeQ[n],
    n=10n+1];
Print[n]
```

```
1111111111111111111
```

!PrimeQ[n] is our Boolean statement. Recall that ! here stands for negation, or Not (see page 104). That is, if a statement, say s, is True, then !s is

False. Here n=10n+1 is the code we want to repeat. The code n=10n+1 simply takes the number n and places 1 at the far right of the number (right?). So the aim is to put as many 1s in front of the original n, which is 111 here, to get a prime number. The While loop does exactly this. It is going to repeat the above code until !PrimeQ[n] becomes False. That is, until PrimeQ[n] becomes True, which happens when n becomes prime. And this is what we are looking for.

Here is a little test to show that the result of the above code is consistent with the code we wrote on page 83.

```
IntegerDigits[n]
{1,1,1,1,1,1,1,1,1,1,1,1,1,1,1,1,1,1,1}

Length[%]
19
```

⌐▬▬▬▬▬▬▬

▬▬▬ Problem 8.6

Find the smallest positive integer m such that $529^3 + 132^3m$ is divisible by 262417.

⟹ SOLUTION.

We start with $m = 1$ and while the remainder on division of $529^3 + 132^3m$ by 262417 is not zero, in *Mathematica* terms, While[Mod[529^ 3 + 132^ 3m, 262417] != 0, we add one to m, i.e., m++, and repeat this until the remainder is zero. Then this is the m we are looking for:

```
m = 1;
While[Mod[529^3 + 132^3 m , 262417] != 0, m++]; m
1984
```

Note that m++ is equivalent to m=m+1, which adds one to m.

One can also use Select in the spirit of Problem 4.17 to get the result (if one knows a bound for m) as follows:

```
Select[Range[10000], Mod[529^3 + 132^3 # , 262417] == 0 &, 1]
{1984}
```

⌐▬▬▬▬▬▬▬

▬▬▬ Problem 8.7

Find the closest prime number less than a given number n.

$\Longrightarrow$ SOLUTION.

Here we have an example which can be "naturally" written by `While`. Here is our first attempt.

```
n = Input["enter a number"]
While[! PrimeQ[n];
    n--];
Print[n]
```

`Input` opens a box and asks for a value. This is a good approach if one wants to ask a user for data. If we enter 6 as an input, we get 5 as an answer, which is the closest prime number less than 6. However, if we enter 111 we get 110, which definitely is not a prime number. A closer inspection shows that the condition `! PrimeQ[n]` is not separated from the body of the code by a comma, rather by a semicolon, which makes it effectively part of the body of the code. Correcting it, we have,

```
n = Input["enter a number"]
While[! PrimeQ[n],
    n--];
Print[n]
```

Notice that the body of the loop contains one line. Now entering 111 returns 109 which is a prime number, i.e., `n--`. Again `!PrimeQ[n]` returns `True` and keeps the loop repeating until `n` is prime. That's what the question asks.

Problem 8.8

Find all prime numbers less than a given n.

$\Longrightarrow$ SOLUTION.

We will use the loop `While` to find one by one all the prime numbers smaller than n starting from the smallest prime number 2. Notice that here the body of `While` has two sentences.

```
i = 1; n = Input["enter a number"]; pset = {};
While[Prime[i] ≤ n,
    pset = pset ∪ {Prime[i]};
    i++];
pset
```

Ok, for $n = 321$ we get all the prime numbers up to 321.

{2, 3, 5, 7, 11, 13, 17, 19, 23, 29, 31, 37, 41, 43, 47, 53, 59, 61, 67, 71, 73, 79, 83, 89, 97, 101, 103, 107, 109, 113, 127, 131, 137, 139, 149, 151, 157, 163, 167, 173, 179, 181, 191, 193, 197, 199, 211, 223, 227, 229, 233, 239, 241, 251, 257, 263, 269, 271, 277, 281, 283, 293, 307, 311, 313, 317}

Here, until `Prime[i]` is smaller than `n`, the loop keeps collecting `Prime[i]` in a list `pset` which at the beginning we set empty (see Example 6.3). After each step we go a step forward by adding one to `i`, that is `i++`, and repeat the same procedure again until `Prime[i]` is bigger than `n`.

Note that one could use `AppendTo` instead of `pset = pset ∪ {Prime[i]}` as follows: `AppendTo[pset,Prime[i]]` (see page 108 for more discussion on `AppendTo`).

Exercise 8.2

Find the smallest positive multiple of 99999 that contains no 9's amongst its digits. (Hint, see `MemberQ`.)

_____ Problem 8.9

Determine the highest power of 5 that divides $\dfrac{(5n)!}{(n!)^5}$ for $1 \le n \le 200$.

⟹ SOLUTION.

```
A = {}; Do[i = 0;
 While[Mod[(5 n)!/(n!)^5, 5^i] == 0,
  AppendTo[A, {n, i}]; i = i + 1],
 {n, 1, 200}];

Max[Last /@ A]
12

Select[A, #[[2]] == 12 &]
{{124, 12}}
```

The result shows that, for $n = 124$, 5^{12} divides $\frac{(5n)!}{(n!)^5}$ and this is the highest power. The code consists of two loops, one `Do` loop to change the value of n and a `While` loop inside it to determine what powers of 5 divide $\frac{(5n)!}{(n!)^5}$. This is an example of a nested loop that we will study in Section 8.2.

The last loop in _Mathematica_ is the `For` loop. Here is the easiest example:

```
For[i=5,i<10,i++,Print[i]]
```

```
5
6
7
8
9
```

The loop `For` consists of different parts as follows:

$$For[init,condition,steps,body].$$

The `init` part is where we initialise the variables we need to use in the body of the loop. In the above example this was `i=5`. The second part is where a Boolean expression appears and is where we decide when to terminate the loop. The last part is reserved for the body of the loop. Each of these parts can have several sentences which should be separated by `;`. Let us look at another example.

▬▬ Problem 8.10

Find the sum of the sequence

$$\frac{1}{1+2} + \frac{2}{2+3} + \cdots + \frac{10}{10+11}.$$

⟹ SOLUTION.

```
For[i = 1; sum = 0, i < 11, i++, sum += i/(i + i + 1)];
sum
64157087/14549535
```

Notice that the `init` part of the loop consists of two lines. Also notice that `sum+=i/(i + i + 1)` is a shorthand for `sum=sum+i/(i + i + 1)` as `i++` is a shorthand for `i=i+1`. In the same way `i*=n` is a shorthand for `i=i*n`.

To refresh our memory, here are the other approaches used to get the sum of the above sequence

```
Sum[i/(2i + 1), {i, 1, 10}]
64157087/14549535
```

```
Plus @@ (#/(2# + 1) & /@ Range[10])
64157087/14549535
```

One can leave out any part of a `For` loop. For example

```
For[ ,False , , Print["Never see the light of day"]]
```

produces nothing. One can also see that `While[test,body]` is the same as `For[ ,test, , body]` and `Do[body,{x,xmin,xmax,inc}]` is the same as `For[x=xmin,x≤xmax,x+=inc,body]`. But again, there are times when `While` makes the code more readable and there are times when `For` is a better choice.

Let us do some experiments:

```
Timing[Do[,{10^6}]]
{0.02 Second,Null}

Timing[Do[,{1000000}]]
{0.1 Second,Null}

Timing[i=1;While[i<10^6,i++]]
{2.614 Second,Null}

Timing[i=1;While[i<1000000,i++]]
{1.932 Second,Null}

Timing[For[i=1,i<10^6,i++]]
{2.654 Second,Null}

Timing[For[i=1,i<1000000,i++]]
{1.973 Second,Null}
```

Here is one more example.

────── Problem 8.11

An integer $d_n d_{n-1} d_{n-2} \ldots d_1$ is *palindromic* if

$$d_n d_{n-1} d_{n-2} \ldots d_1 = d_1 d_2 \ldots d_{n-1} d_n$$

(for example 15651). Write a code to ask for a number $d_n d_{n-1} d_{n-2} \ldots d_1$ and find out if it is palindromic. Enhance the code further so that if the number is not palindromic then the code tests whether $d_n d_{n-1} d_{n-2} \ldots d_1 + d_1 d_2 \ldots d_{n-1} d_n$ is (for example, 108+801=909). Furthermore, write a code to give the number of times it is needed to repeat this procedure until one gets a palindromic number starting with $d_n d_{n-1} d_{n-2} \ldots d_1$ (if it takes more than 150 times, let the function return infinity).

⟹ SOLUTION.

We start with an example. Let $n = 98$. We need to check systematically whether n is palindromic. If not, then produce 89, add this to $n = 98$ and check whether this is palindromic. We have seen how to produce the reverse of a number using `IntegerDigits, Reverse` and `FromDigits` (see Problem 4.13). Here is the first step:

```
n=98;nlist=IntegerDigits[n]
{9,8}
```

```
If[nlist != Reverse[nlist],n=n+FromDigits[Reverse[nlist]]]
187
```

If the result is not palindromic, one has to do the same procedure again. Thus we use a `While` loop to do this for us.

```
n = 98; nlist = IntegerDigits[n];
While[nlist != Reverse[nlist],
  n = n + FromDigits[Reverse[nlist]];
  nlist = IntegerDigits[n]
  ]; n
```

```
8813200023188
```

One can enhance the code:

```
n = Input["Enter a number"]; i = 1; nlist = IntegerDigits[n];
safetyNet = True;

While[nlist != Reverse[nlist] && safetyNet,
  Print[i, " ", n]; n = n + FromDigits[Reverse[nlist]]; i++;
  nlist = IntegerDigits[n];
  If[i > 150, safetyNet = False]
  ]
If[i > 150, Print[".........Aborted"], Print[i, " ", n]
  ]
```

8.2 Nested loops

In many applications there are several factors (variables) which change simultaneously, and this calls for what we call a *nested loop*. Instead of trying to describe the situation, let us look at some examples.

```
Do[
  Do[
    Print[i, " ", j],
    {j, 1, 2}
    ],
  {i, 1, 3}
  ]
```

```
1 1
1 2
2 1
2 2
3 1
3 2
```

The code contains two Do loops, one inside the other. In the inner one, the counter j runs from 1 to 2 and once this is done, the outer loop performs and the counter i goes one further and again the inner loop starts to run.

───── Problem 8.12

Find all the pairs (n, m) for $n, m \leq 10$ such that $n^2 + m^2$ is a square number (e.g., $(3, 4)$ as $3^2 + 4^2 = 5^2$).

⟹ SOLUTION.

```
Do[
  Do[
    If[Sqrt[i^2 + j^2] ∈ Integers, Print[i, " ",
  j]],
    {j, i, 10}
    ],
  {i, 1, 10}
  ]
```

Here is the result

```
3  4
6  8
```

Here the outer loop starts with the counter i getting the value 1. Then it is the turn of the block inside this loop, which is again another loop run. In the inner loop {j, i, 10} makes the counter j run from i to 10. This done, in the outer loop i takes 2 and then j runs from 2 to 10 and so on and each time checks whether $\sqrt{i^2 + j^2}$ is an integer. The reader should see that this is enough to find all the pairs up to 10 with the desired property. Can you say how many times the If line is going to be performed?

One can make the nested Do loop a bit shorter. The following is an equivalent code to the first example of a Nested Do loop on page 132.

```
Do[
 Print[i , " ", j],
 {i, 1, 3}, {j, 1, 2}]

1 1
1 2
2 1
2 2
3 1
3 2
```

Note that here j is the counter for the inner loop.

We have already seen the command `Table` which provides a sort of loop. In fact, `Table` can provide us with a nested loop as well.

```
Table[{i, j}, {i, 1, 3}, {j, 1, 2}]
{{{1, 1}, {1, 2}}, {{2, 1}, {2, 2}}, {{3, 1}, {3, 2}}}
```

One should compare this with the example of the nested `Do` loop above. As the result shows, here j is the counter for the inner loop. One of the issues that might arise here is that the output is a nested list (i.e., too many { and }). Sometimes we really do not need the nested list answer to our question. For example, we want to come up with a code to solve Problem 8.12 by using `Table`. In order to get rid of extra "{", one can use the command `Flatten`.

```
Flatten[Table[{i, j}, {i, 1, 3}, {j, 1, 4}]]
{1, 1, 1, 2, 1, 3, 1, 4, 2, 1, 2, 2, 2, 3, 2, 4, 3, 1, 3, 2, 3,
3, 3, 4}
```

`Flatten` gets rid of all the lists inside a list, i.e., removes all the "{". In our problem we want a list of pairs. In this case we need

```
Flatten[Table[{i, j}, {i, 1, 3}, {j, 1, 4}],1]
{{1, 1}, {1, 2}, {1, 3}, {1, 4}, {2, 1}, {2, 2}, {2, 3}, {2, 4},
{3, 1}, {3, 2}, {3, 3}, {3, 4}}
```

Now we have all the pairs. But some of them are repeated. For us $\{1,3\}$ is the same as $\{3,1\}$. So, as in Problem 8.12, we need the inner counter to depend on the outer one as follows:

```
Flatten[Table[{i, j}, {i, 1, 3}, {j, i, 4}], 1]
{{1, 1}, {1, 2}, {1, 3}, {1, 4}, {2, 2}, {2, 3}, {2, 4}, {3, 3},
{3, 4}}
```

All we have to do now is to select the pairs (m, n) such that $\sqrt{m^2 + n^2} \in \mathbb{N}$.

```
Select[Flatten[Table[{i, j}, {i, 1, 10}, {j, i,
10}], 1],
(Sqrt[#[[1]]]^2 + #[[2]]^2] ∈ Integers) &]


{{3, 4}, {6, 8}}
```

Exercise 8.3

A pair (m, n) such that $m^2 + n^2$ is a square number is called a Pythagorean pair (see Problem 8.12). Find a Pythagorean pair (m, n) such that if the digits of m are written in reverse order, then n is obtained.

Exercise 8.4

Pick an odd prime number p. Then find a pair (q, r) of positive integers such that $p^2 + q^2 = r^2$.

8.3 Nest, NestList and more

Let $f(x)$ be a function defined on a variable x. There are times when one needs to apply the function f to itself several times, i.e., $f(\cdots f(f(x))\cdots)$ (see the examples on page 51). *Mathematica* provides a command to do exactly this:

```
Nest[f, x, 4]
f[f[f[f[x]]]]
```

If one wants to keep track of each step, the command `NestList` is available

```
NestList[f, x, 4]
{x, f[x], f[f[x]], f[f[f[x]]], f[f[f[f[x]]]]}
```

Here are some (nice) examples:

```
f[x_]:=1/(1+x)
Nest[f,x,4]
```

$$\cfrac{1}{1+\cfrac{1}{1+\cfrac{1}{1+\cfrac{1}{1+x}}}}$$

```
NestList[f,x,4]
```

$$\left\{\frac{1}{1+x}, \cfrac{1}{1+\frac{1}{1+x}}, \cfrac{1}{1+\cfrac{1}{1+\frac{1}{1+x}}}, \cfrac{1}{1+\cfrac{1}{1+\cfrac{1}{1+\frac{1}{1+x}}}}\right\}$$

```
NestList[Sqrt[#+6]&,Sqrt[6],4]
```

$$\{\sqrt{6}, \sqrt{6 + \sqrt{6}}, \sqrt{6 + \sqrt{6 + \sqrt{6}}}, \sqrt{6 + \sqrt{6 + \sqrt{6 + \sqrt{6}}}},$$

$$\sqrt{6 + \sqrt{6 + \sqrt{6 + \sqrt{6 + \sqrt{6}}}}}\}$$

Using the dynamic variable n we can monitor how the continuous function "grows" as n runs from 1 to 10.

```
f[x_] := 1 / (1 + x)
```

```
Manipulate[NestList[f, x, n], {n, 1, 10, 1}]
```

Exercise 8.5

Find the roots of the equation

$$x = 1 + \cfrac{1}{1 + \cfrac{1}{1 + \cfrac{1}{1 + \cdots \frac{1}{1+x}}}}$$

where there are 10 division lines in the expression on the right. (Hint: use Solve to find roots, more on this in Chapter 15.)

Exercise 8.6

Investigate that

$$\frac{2}{\pi} = \frac{\sqrt{2}}{2} \frac{\sqrt{2 + \sqrt{2}}}{2} \frac{\sqrt{2 + \sqrt{2 + \sqrt{2}}}}{2} \cdots$$

There are two more commands of this type, NestWhile and NestWhileList.

> ?NestWhile
> NestWhile[f, expr, test] starts with expr, then
> repeatedly applies f until applying test to the
> result no longer yields True.

The following problem uses NestWhile.

_____ Problem 8.13

A happy number is a number such that if one squares its digits and adds them together, and then takes the result and squares its digits and adds them together again and keeps doing this process, one comes down to the number 1. Find all the happy ages, i.e., happy numbers up to 100.

$\implies$ SOLUTION.

```
Select[Range[100],
   NestWhile[
      Plus @@ (IntegerDigits[#]^2)&,#,(!#==4) && (!#==1)&]==1&]
```

$\{1,7,10,13,19,23,28,31,32,44,49,68,70,79,82,86,91,94,97,100\}$

Deciphering this code is a bit challenging. Note that the code contains three pure functions (so, three & for those) and a boolean expression containing And (so && for And).

There is a more elegant approach to this problem using recursive functions in Problem 12.5. In any case, one can observe that happy ages are mostly before one gets a job or after retirement!

Now, we are going to make up a problem and use NestList to get some answers.

_____ Problem 8.14

A number $a_1 a_2 \cdots a_n$ is called a pure prime number if

$$a_1 a_2 \ldots a_n, \; a_1 a_2 \ldots a_{n-1}, \ldots, a_1 a_2 a_3, \; a_1 a_2, \; \text{and} \; a_1$$

are all prime. Prove that pure prime numbers are finite in number and find all
of them.

⟹ SOLUTION.

First we have to find a way to drop the last digit of a number. The function
Quotient might help

```
?Quotient

Quotient[m, n] gives the integer quotient of m and n.

Quotient[5937, 10]
593

Quotient[593, 10]
59

Quotient[59 , 10]
5
```

The above example shows that applying Quotient to a number repeatedly
drops the last digit of the number one by one. Thus

```
NestList[Quotient[#, 10] &, 5937, 3]

{5937,593,59,5}
```

Now we have all the numbers. We only need to test whether all of them are
prime.

```
PrimeQ /@ NestList[Quotient[#,10]&,5937,3]
{False, True, True, True}
```

Thus 5937 just misses being a pure prime. If we want to define this as
a function, a little problem might arise. In the case of 5937 we have to apply
Quotient three times to this number. But for a number n with arbitrary digits,
we need to use FixedPointList.

```
?FixedPointList

FixedPointList[f, expr] generates a list giving the results of
applying f  repeatedly, starting with expr, until the results no
longer change.

FixedPointList[Quotient[#,10]&,5937]
{5937, 593, 59, 5, 0, 0}

FixedPointList[Quotient[#,10]&,7647653]
{7647653, 764765, 76476, 7647, 764, 76, 7, 0, 0}
```

It is clear that we have to drop the last two elements from the list.

```
Drop[FixedPointList[Quotient[#,10]&,5937],-2]
{5937, 593, 59, 5}
```

Now it is time to apply `PrimeQ` to the list to check whether all these numbers are prime.

```
PrimeQ[Drop[FixedPointList[Quotient[#,10]&,5937],-2]]
{False,True,True,True}
```

What we need is a list containing only `True`'s. Thus if only one of the numbers happens to be not prime, the whole number is not pure prime, as is the case with 5937. We can combine all the Booleans in the list with `And` and the result would make it clear whether the number is pure prime. Here is the code

```
Apply[And,{False,True,True,True}]
False
```

Thus, putting all of this together, we have

```
purePrime[n_]:=Apply[And,PrimeQ[Drop[FixedPointList[
Quotient[#,10]&,n],-2]]]

Select[Range[10, 99], purePrime]
{23,29,31,37,53,59,71,73,79}

Select[Range[100,999],purePrime]
{233, 239, 293, 311, 313, 317, 373, 379, 593, 599, 719, 733, 739,
797}

Select[Range[1000, 9999], purePrime]
{2333,2339,2393,2399,2939,3119,3137,3733,3739,3793,3797,5939,
7193,7331,7333,7393}
```

This seems not to be a good algorithm to find all pure prime numbers as it already takes some time to find all the 6-digit pure primes. In fact the problem here is that in order to find, say, all the 4-digit pure primes, the above algorithm has to check all the numbers from 1000 to 9999. But this is not necessary. The following example demonstrates this. If we know that 719 is pure prime then all we have to check, to find the pure primes which have four digits and whose first three digits are 719, are the numbers $\{7190, 7191, ..., 7199\}$.

```
Range[719*10, 719*10 + 9]
{7190,7191,7192,7193,7194,7195,7196,7197,7198,7199}
```

We do not need to consider even numbers.

```
Range[719*10+1, 719*10 + 9,2]
{7191,7193,7195,7197,7199}
```

Now we need to find out which of these numbers are prime.

```
Select[%, PrimeQ]
{7193}
```

This shows that 7193 is a pure prime. Thus we start with all one-digit primes and find all the two-digit pure primes as above.

```
purelist={2,3,5,7}
{2, 3, 5, 7}
```

```
Range[10#+1,10#+9,2]&[purelist]
{{21, 23, 25, 27, 29}, {31, 33, 35, 37, 39}, {51, 53, 55, 57, 59},
{71, 73, 75, 77, 79}}
```

```
Flatten[Range[10#+1,10#+9,2]&[purelist]]
{21, 23, 25, 27, 29, 31, 33, 35, 37, 39, 51, 53, 55, 57, 59, 71,
73, 75, 77, 79}
```

```
purelist=Select[Flatten[Range[10#+1,10#+9,2]&[purelist]],PrimeQ]
{23, 29, 31, 37, 53, 59, 71, 73, 79}
```

Thus these are all the two-digit pure primes. Now that we have found them, we can immediately find all three-digit primes.

```
purelist=Select[Flatten[Range[10#+1,10#+9,2]&[purelist]],PrimeQ]
{233,239,293,311,313,317,373,379,593,599,719,733,739,797}
```

Thus our clever code to find all the pure primes is as follows:

```
purelist={2,3,5,7};
While[purelist != {},
  purelist=Select[Flatten[Range[10#+1,10#+9,2]&[purelist]],PrimeQ];
  Print[purelist]
  ]
```

```
{23,29,31,37,53,59,71,73,79}
```

```
{233,239,293,311,313,317,373,379,593,599,719,733,739,797}
```

```
{2333,2339,2393,2399,2939,3119,3137,3733,3739,3793,3797,5939,
7193,7331,7333,7393}
```

```
{23333,23339,23399,23993,29399,31193,31379,37337,37339,37397,
59393,59399,71933,73331,73939}
```

```
{233993,239933,293999,373379,373393,593933,593993,719333,739391,
739393,739397,739399}
```

```
{2339933,2399333,2939999,3733799,5939333,7393913,7393931,7393933}
```

```
{23399339,29399999,37337999,59393339,73939133}
```

♣ TIPS

– In Problem 8.14 we used the command FixedPointList. There are also two other commands in the same spirit, namely, LengthWhile and TakeWhile.

Exercise 8.7

Starting with a number, consider the sum of all the proper divisors of the number. Now consider the sum of all the proper divisors of this new number and repeat this process. If one eventually obtains the number which one started with, then this number is called a *social number*. Write a program to show that 1264460 is a social number.

8.4 Fold and FoldList

Recall one of the questions we asked in Chapter 4, namely: Given $\{x_1, x_2, \cdots, x_n\}$ how can one produce $\{x_1, x_1 + x_2, \cdots, x_1 + x_2 + \cdots + x_n\}$?

Let us look at the commands Fold and FoldList.

```
Fold[f,x,{a,b,c}]
f[f[f[x,a],b],c]
```

```
FoldList[f,x,{a,b,c}]
{x,f[x,a],f[f[x,a],b],f[f[f[x,a],b],c]}
```

Replace the function f with Plus and x with 0 and observe what happens. (See the following problem for the answer.) Here is a use of FoldList to write another code for Problem 7.3.

━━━ Problem 8.15

Write a function to calculate the sum of the following sequence.

$$p(n) = \frac{1}{1} + \frac{1}{1+2} + \dots + \frac{1}{1+2+\dots+n}$$

⟹ SOLUTION.

Here is the code:

```
p[n_]:= Plus @@ (1/Rest[FoldList[Plus, 0, Range[n]]])
```

In order to decipher this code, let us look at the standard example of FoldList.

```
FoldList[Plus,0,{a,b,c}]
{0,a,a+b,a+b+c}
```

Thus, dropping the annoying 0 from the list:

```
Rest[FoldList[Plus,0,{a,b,c}]]
{a,a+b,a+b+c}
```

and

```
1/Rest[FoldList[Plus,0,{a,b,c}]]
```

$$\left\{\frac{1}{a}, \frac{1}{a+b}, \frac{1}{a+b+c}\right\}$$

makes the original code clear.

We have used `FoldList` in its simplest form, that is,

```
FoldList[Plus,0,{a,b,c}]
{0,a,a+b,a+b+c}
```

If we are just after this, namely, "a list of the successive accumulated totals of elements in a list", then the command `Accumulate` does just that as well:

```
Accumulate[{a, b, c, d}]
{a, a + b, a + b + c, a + b + c + d}
```

Later, in Problems 14.8 and 14.10, we will use these commands to generate very interesting graphs.

Here is one more example where the function used in `FoldList` is not `Plus`.

```
Manipulate[FoldList[Sqrt[#1 + #2] &, x, Range[i]], {i, 1, 10, 1}]
```

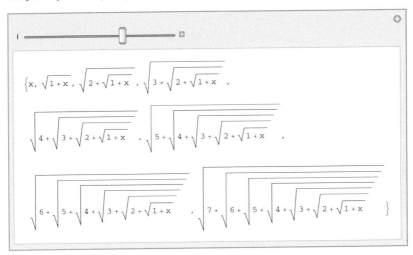

8.5 Inner and Outer

Recall the following questions from Chapter 4: Given $\{x_1, x_2, \cdots, x_n\}$ and $\{y_1, y_2, \cdots, y_n\}$, how can one produce the combinations

$$\{x_1, y_1, x_2, y_2, x_3, y_3, \cdots, x_n, y_n\},$$

$$\{\{x_1, y_1\}, \{x_2, y_2\}, \cdots, \{x_n, y_n\}\},$$

and

$$\{\{x_1, y_1\}, \{x_1, y_2\}, \cdots, \{x_1, y_n\},$$
$$\{x_2, y_1\}, \{x_2, y_2\}, \cdots, \{x_2, y_n\}, \cdots, \{x_n, y_1\}, \{x_n, y_2\}, \cdots, \{x_n, y_n\}\}?$$

One can get the first two lists using the command `Inner` and the third list by using `Outer`:

```
Inner[f,{a,b},{x,y},g]
g[f[a,x],f[b,y]]
```

If we replace the functions `f` and `g` with `List` we get:

```
Inner[List,{a,b},{x,y},List]
{{a,x},{b,y}}

Flatten[%]
{a,x,b,y}
```

Or this one to get rid of `Flatten`:

```
Inner[Sequence,{a,b},{x,y},List]
{a,x,b,y}
```

So putting these together we can write

```
Inner[List, Map[x#& Range[5]], Map[y#& Range[5]], List]
{{x_1,y_1},{x_2,y_2},{x_3,y_3},{x_4,y_4},{x_5,y_5}}
```

There is a more clever way to get this using `Transpose` (see Chapter 13).

```
Flatten[Transpose[{{a, b}, {x, y}}]]
{a, x, b, y}
```

Now to obtain the second list:

```
Outer[f, {a, b}, {x, y, z}]
{{f[a, x], f[a, y], f[a, z]}, {f[b, x], f[b, y], f[b, z]}}

Outer[List, {a, b}, {x, y, z}]
{{{a, x}, {a, y}, {a, z}}, {{b, x}, {b, y}, {b, z}}}

Flatten[Outer[List, {a, b}, {x, y, z}], 1]
{{a, x}, {a, y}, {a, z}, {b, x}, {b, y}, {b, z}}
```

As one can imagine, commands of this type provide many possibilities and inspiring compositions of functions! (See Exercise 8.8.)

_____ Problem 8.16

Consider the pairs (m, n) where $50 \leq m, n \leq 60$. We want to consider a graph with vertices the numbers 50 to 60 and an *edge* from m to n if mn is a prime number. For example, there is an edge between 50 and 51 as 5051 is a prime number. Draw this graph.

$\Longrightarrow$ SOLUTION.

We first produce all the pairs (m, n) where $1 \leq m, n \leq 30$. This can easily be done with Outer. Here is an example for a smaller range:

```
s = Flatten[Outer[List, Range[10, 15], Range[10, 15]], 1]
```

```
{{10, 10}, {10, 11}, {10, 12}, {10, 13}, {10, 14}, {10,
  15}, {11, 10}, {11, 11}, {11, 12}, {11, 13}, {11, 14}, {11,
  15}, {12, 10}, {12, 11}, {12, 12}, {12, 13}, {12, 14}, {12,
  15}, {13, 10}, {13, 11}, {13, 12}, {13, 13}, {13, 14}, {13,
  15}, {14, 10}, {14, 11}, {14, 12}, {14, 13}, {14, 14}, {14,
  15}, {15, 10}, {15, 11}, {15, 12}, {15, 13}, {15, 14},
  {15, 15}}
```

In Problem 8.17 we define the function $\diamond$ for this Cartesian product.

Next we define a function to get a list of two numbers {m,n} and put them together as mn.

```
fd[n_] := FromDigits[Flatten[IntegerDigits[n], 1]]
```

```
fd[{23, 76}]
2376
```

Now we are ready to put these together.

```
s = Flatten[Outer[List, Range[50, 60], Range[50, 60]], 1];
```

We consider each pair {m,n}, then, using the function fd, we check if mn is prime. If this is the case we then collect $m \to n$ (using Rule). This is because of how the command GraphPlot works.

```
t = If[PrimeQ[fd[#]], Rule @@ #, 0] & /@ s
```

```
{0, 50 -> 51, 0, 0, 0, 0, 0, 0, 0, 50 -> 59, 0, 0, 0, 0,
 51 -> 53, 0, 0, 0, 0, 0, 0, 0, 0, 0, 0, 0, 0, 0, 0, 0, 0,
 0, 53 -> 51, 0, 0, 0, 0, 0, 0, 0, 0, 0, 0, 0, 0, 0, 0, 0,
 0, 0, 0, 0, 0, 0, 0, 0, 0, 0, 55 -> 57, 0, 0, 0, 0, 56 -> 51,
 0, 56 -> 53, 0, 0, 0, 56 -> 57, 0, 56 -> 59, 0, 0, 0, 0, 0,
 0, 0, 0, 0, 0, 0, 0, 0, 58 -> 51, 0, 0, 0, 0, 0, 58 -> 57, 0,
 0, 0, 0, 0, 0, 59 -> 53, 0, 0, 0, 0, 0, 0, 0, 0, 0, 60 -> 53,
 0, 0, 0, 0, 0, 0, 0}
```

We remove the zeros from the list.

```
k = DeleteCases[t, 0]
```

```
{50 -> 51, 50 -> 59, 51 -> 53, 53 -> 51, 55 -> 57, 56 -> 51,
 56 -> 53, 56 -> 57, 56 -> 59, 58 -> 51, 58 -> 57, 59 -> 53,
 60 -> 53}
```

We are ready to plot the graph:

```
GraphPlot[k, VertexLabeling -> True, DirectedEdges -> True]
```

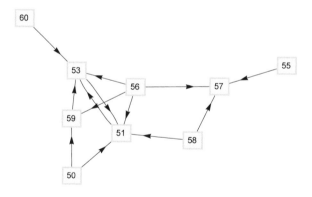

We can assemble these together using Manipulate to observe how the graph changes as we go through different range of numbers.

```
s[j_] := Flatten[
  Outer[List, Range[10*j, 10*j + 9], Range[10*j, 10*j + 9]], 1]
```

```
Manipulate[t = If[PrimeQ[fd[#]], Rule @@ #, 0] & /@ s[j];
 k = DeleteCases[t, 0];
 GraphPlot[k, VertexLabeling -> True, DirectedEdges -> True],
 {j, 5, 100, 10}]
```

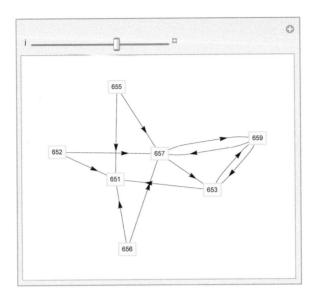

_____

_____ Problem 8.17

We call a pair of prime numbers p and q *friends* if pq and qp are both prime
(by pq we mean the juxtaposition of the numbers together). For example, the
prime numbers 563 and 587 are friends as 563587 and 587563 are both primes.
Write a program to produce all the prime friends for some suitable range of
primes.

$\Longrightarrow$ SOLUTION.

We look at the pairs (n, m), produce the nth and mth prime numbers and
then put these numbers together and check if they are prime.

To consider all the pairs (n, m), we define a Cartesian product of two lists
using Outer as described in this section.

```
Flatten[Outer[List, Range[50, 55], Range[50, 55]], 1]
```

```
{{50, 50}, {50, 51}, {50, 52}, {50, 53}, {50, 54}, {50, 55},
 {51, 50}, {51, 51}, {51, 52}, {51, 53}, {51, 54}, {51, 55},
 {52, 50}, {52, 51}, {52, 52}, {52, 53}, {52, 54}, {52, 55},
 {53, 50}, {53, 51}, {53, 52}, {53, 53}, {53, 54}, {53, 55},
 {54, 50}, {54, 51}, {54, 52}, {54, 53}, {54, 54}, {54, 55},
 {55, 50}, {55, 51}, {55, 52}, {55, 53}, {55, 54}, {55, 55}}
```

To improve the readability, we define this as a function as follows:

```
a_ ◇ b_ := Flatten[Outer[List, a, b], 1]

Range[5, 8] ◇ Range[-5, -2]

{{5, -5}, {5, -4}, {5, -3}, {5, -2}, {6, -5}, {6, -4}, {6, -3},
{6, -2}, {7, -5}, {7, -4}, {7, -3}, {7, -2}, {8, -5}, {8, -4},
{8, -3}, {8, -2}}
```

We have seen functions of this type in Chapter 3, where we defined a function called $\oplus$. Here the function is represented by $\diamond$ (see Problem 10.10 for a detailed explanation). The symbol $\diamond$ can be typed by pressing the keys "esc" dia "esc".

Next we define a function that puts two numbers next to each other. We represent the function by $\triangleleft$.

```
x_Integer ◁ y_Integer := FromDigits[
  IntegerDigits[x] ~ Join ~ IntegerDigits[y]]

124 ◁ 652


124652
```

Now we are ready to use these functions:

```
t1 = Select[Prime[Range[100, 110] ◇ Range[100, 110]],
  PrimeQ[#[[1]] ◁ #[[2]]] &]

t2 = Reverse /@ t1
{{547, 541}, {571, 541}, {577, 541}, {577, 547}, {601,
  547}, {587, 563}, {593, 563}, {599, 563}, {599, 569}, {541,
  571}, {601, 571}, {547, 577}, {601, 577}, {563, 587}, {599,
  587}, {587, 593}, {541, 601}}

t1 \[Intersection] t2

{{541, 571}, {547, 577}, {563, 587}, {571, 541}, {577, 547},
  {587, 563}}

PrimeQ[563587]
True

PrimeQ[587563]
True
```

Exercise 8.8

Write a code to convert $\{x_1, x_2, \ldots, x_k, x_{k+1}\}$ to

$$\{x_1 \to x_2, x_2 \to x_3, \ldots, x_k \to x_{k+1}, x_{k+1} \to x_1\}.$$

Exercise 8.9

Consider a positive integer. Then sort the decimal digits of this number in ascending and descending order. Subtract these two numbers (for example, starting from 5742, we get $7542 - 2457 = 5085$). This is called the Kaprekar routine. First check that starting with any 4-digit number and repeating the Kaprekar routine, you always reach either 0 or 6174. Then find out, among all the 4–digit numbers, what is the maximum number of iterations needed in order to get 6174.

9

Substitution, Mathematica rules

This chapter introduces a way to substitute an expression by another expression without changing their values. This is done by setting a substitution rule for *Mathematica* to follow. This simple idea provides a way to write very elegant programs.

In Wolfram *Mathematica*® one can substitute an expression with another using *rules*. In particular, one can substitute a variable with a value without assigning the value to the variable. Here is how it goes:

```
x + y /. x -> 2
2 + y
```

This means we replace x with 2 without assigning 2 to x. If we ask for the value of x, we see

```
x
x
```

```
FullForm[x + y /. x -> 2]
Plus[2,y]
```

The following examples show the variety of things one can do with rules.

```
x + y /. {x -> a, y -> b}
a + b
```

```
x^2 - 3 /. x -> Sqrt[3]
0
```

```
x^2 + y /. x -> y /. y -> x
x + x^2
```

When solving an equation using `Solve` the output is in the form of rules.

© Springer International Publishing Switzerland 2015

R. Hazrat, *Mathematica®: A Problem-Centered Approach*, Springer Undergraduate Mathematics Series, DOI 10.1007/978-3-319-27585-7_9

```
Solve[x^3 - 2 x + 1 == 0]
```

$$\left\{ \{x \to 1\}, \left\{ x \to \frac{1}{2}\left(-1 - \sqrt{5}\right) \right\}, \left\{ x \to \frac{1}{2}\left(-1 + \sqrt{5}\right) \right\} \right\}$$

```
x /. %
```

$$\left\{ 1, \frac{1}{2}\left(-1 - \sqrt{5}\right), \frac{1}{2}\left(-1 + \sqrt{5}\right) \right\}$$

```
s = Solve[x^3 + x^2 + x + 6 == 0, x]
```

$$\left\{ \{x \to -2\}, \left\{ x \to \frac{1}{2}\left(1 - i\sqrt{11}\right) \right\}, \left\{ x \to \frac{1}{2}\left(1 + i\sqrt{11}\right) \right\} \right\}$$

```
x^3 + x^2 + x + 6 /. s
```

$$\left\{ 0, 6 + \frac{1}{2}\left(1 - i\sqrt{11}\right) + \frac{1}{4}\left(1 - i\sqrt{11}\right)^2 + \frac{1}{8}\left(1 - i\sqrt{11}\right)^3, \right.$$
$$\left. 6 + \frac{1}{2}\left(1 + i\sqrt{11}\right) + \frac{1}{4}\left(1 + i\sqrt{11}\right)^2 + \frac{1}{8}\left(1 + i\sqrt{11}\right)^3 \right\}$$

```
Simplify[x^3 + x^2 + x + 6 /. s]
```

$\{0, 0, 0\}$

Here is an amusing way to change one pattern into another.

```
Range[5] * x
{x, 2 x, 3 x, 4 x, 5 x}

Range[5] * x /. Times -> Power
{x, 2^x, 3^x, 4^x, 5^x}
```

Looking at the fullform of this expression, it makes it clear that when we replace Times by Power we obtain the above expression.

```
FullForm[Range[5]*x]

List[x,Times[2,x], Times[3,x], Times[4,x], Times[5,x]]
```

Let us look at some other expressions.

```
x + 2 y /. {x -> y, y -> z}
y+2z

FullForm[x + 2 y /. {x -> y, y -> z}]
Plus[y, Times[2, z]]
```

This example reveals that *Mathematica* goes through the expression only once and replaces the rules. If we need *Mathematica* to go through the expression again and repeat the whole process until no further substitution is possible, one uses //. as follows:

```
x + 2 y //. {x -> y, y -> z}
3 z
```

In fact /. and //. are shorthand for ReplaceAll and ReplaceRepeated respectively.

Be careful not to confuse both *Mathematica* and yourself:

```
ReplaceRepeated[x + 2 y, {x -> y, y -> x}]

During evaluation of In[11]:= ReplaceRepeated::rrlim: Exiting
after x+2 y scanned 65536 times. >>
x + 2 y
```

One can use MaxIterations to instruct *Mathematica* how many times to repeat the substitution process. This comes in quite handy in many of the problems we look at in this book:

```
ReplaceRepeated[1/(1 + x), x -> 1/(1 + x), MaxIterations -> 4]

During evaluation of In[14]:= ReplaceRepeated::rrlim: Exiting
after 1/(1+x) scanned 4 times. >>
```

$$\cfrac{1}{1+\cfrac{1}{1+\cfrac{1}{1+\frac{1}{1+x}}}}$$

If you don't want to get the annoying message of "…. scanned 4 times", use Quiet to prevent messages of this kind.

```
Quiet[ReplaceRepeated[1/(1 + x), x -> 1/(1 + x),
MaxIterations -> 4]]
```

$$\cfrac{1}{1+\cfrac{1}{1+\cfrac{1}{1+\frac{1}{1+x}}}}$$

In Chapter 10 we will use rules effectively with the pattern matching facility of *Mathematica* (see, for example, Problem 10.6).

——— Problem 9.1

Generate the following list $\{x^{y^z}, x^{z^y}, y^{x^z}, y^{z^x}, z^{x^y}, z^{y^x}\}$.

⟹ SOLUTION.

The command Permutations generates a list of all possible permutations of the elements. Once we get all the possible arrangements, we can use a rule to change $\{x, y, z\}$ to x^{y^z} as shown below:

```
Permutations[{x, y, z}]
{{x, y, z}, {x, z, y}, {y, x, z}, {y, z, x}, {z, x, y}, {z,
  y, x}}

Permutations[{x, y, z}] /. {a_, b_, c_} -> a^b^c
{x^y^z, x^z^y, y^x^z, y^z^x, z^x^y, z^y^x}
```

Exercise 9.1

Explain what the following code does:

```
(a + b)^n /. (x_ + y_)^z_ -> (x^z + y^z)
```

Problem 9.2

To get the Thue–Morse sequence, start with 0 and then repeatedly replace 0 with 01 and 1 with 10. So the first four numbers in the sequence are

$$0$$
$$01$$
$$0110$$
$$01101001$$

Write a function to produce the n-th element of this sequence. (We will visit this sequence again in Problem 14.10.)

$\Longrightarrow$ SOLUTION.

This problem is just waiting for the *Mathematica* rules to get into action! We define a general rule to replace 0 with 01 and replace 1 with 10. However, to do so, we consider 0 and 1 in a list, and define the rule as 0->{0,1} and 1->{1,0} and, at the end, we get rid of all the parentheses with Flatten and put all the digits together again using FromDigits, as the following codes show:

```
{0} /. {0 -> {0, 1}, 1 -> {1, 0}}
{{0, 1}}

{{0, 1}} /. {0 -> {0, 1}, 1 -> {1, 0}}
{{{0, 1}, {1, 0}}}

{{{0, 1}, {1, 0}}} /. {0 -> {0, 1}, 1 -> {1, 0}}
{{{{0, 1}, {1, 0}}, {{1, 0}, {0, 1}}}}

Flatten[%]
{0, 1, 1, 0, 1, 0, 0, 1}

FromDigits[%]
1101001
```

So we define a function, repeat the above process using ReplaceRepeated and control the number of the iterations using MaxIterations, as in the example on page 151.

```
morse[n_] :=
 Quiet[FromDigits[
  Flatten[ReplaceRepeated[{0}, {0 -> {0, 1}, 1 -> {1, 0}},
   MaxIterations -> n]]]]
```

```
morse[4]
110100110010110
```

```
morse[6]
110100110010110100101100110100110010110011010010110
100110010110
```

Note that *Mathematica* drops the first 0 that Thue–Morse starts with (as she considers this as a number). To avoid this, one can replace 0 with x and 1 with y and then the replacement rules are x->{x,y} and y->{y,x}.

To see how quickly this series grows, wrap the code with Manipulate and run n from 1 to 20 as follows: (Note that we have used Quiet as on page 151 to prevent any messages being produced because of the use of MaxIterations.)

_____ Problem 9.3

For which natural numbers n is it possible to choose signs $+$ and $-$ in the expression

$$1^2 \pm 2^2 \pm 3^2 \pm \cdots \pm n^2$$

so that the result is 0?

$\Longrightarrow$ SOLUTION.

One can find the following code in Vardi [6].

```
Fold[(#1/.x→x+#2)(#1/.x→x-#2)&,x,{a,b,c}]/.x→1
```

```
(1-a-b-c) (1+a-b-c) (1-a+b-c) (1+a+b-c) (1-a-b+c)
(1+a-b+c) (1-a+b+c) (1+a+b+c)
```

Motivated by this code, one can approach the problem as follows.

```
Do[ If[(Fold[(#1/.x→x+#2)(#1/.x→x-#2)&,x,
Range[n]^2]/.x→0) = 0,Print[n] ],
{n,1,40}]
7 8 11 12 15 16 19 20 23
$ Aborted
```

However, this seems to take time and there might be a better way to approach this problem.

Here is another approach:

```
t[n_]:=Flatten[Outer[List,Sequence @@Table[{I
k,k}^2,{k,2,n}]],n-2]

Do[
  If[Select[t[n],Plus @@ #==-1&,1]!={},Print[n]],
  {n,3,40}]

7
8
11
12
15
16
19
20

Hold[Abort[],Abort[]]
```

Is there a better way to do this?

10
Pattern matching

Everything in *Mathematica* is an expression and each expression has a pattern. One can search for a specific pattern and change it to another pattern. This is called pattern matching programming. This chapter explains this method. Get ready to be amazed!

Everything in Wolfram *Mathematica*® is an expression and each expression has a pattern. *Mathematica* provides us with the ability to decide whether an expression matches a specific pattern. Following R. Gaylord's exposition [2], consider the expression x^2. This expression is precisely of the following form or *pattern*, "x raised to the power of two"

```
MatchQ[x^2, x^2]
True
```

But x^2 will be matched also by the following loose description, "something" or "an expression"

```
MatchQ[x^2, _]
True
```

Here _ stands (or rather sits) for an expression (_ is called *a blank* here). Also x^2 will match "x to the power of something"

```
MatchQ[x^2, x^_]
True
```

Before we go further, we need to mention that one can give a name to a blank expression as follows n_. Here the expression _ is labelled n. (We have already seen n_ in defining a function in Chapter 3. In fact, when defining a function, we label an expression that we plug into the function.) One can also restrict the expression by limiting its head! Namely, _head matches an expression with the head head. Look:

© Springer International Publishing Switzerland 2015
R. Hazrat, *Mathematica®: A Problem-Centered Approach*, Springer Undergraduate
Mathematics Series, DOI 10.1007/978-3-319-27585-7_10

```
FullForm[x^2]
Power[x, 2]

Head[x^2]
Power

MatchQ[x^2, _Power]
True

Head[4]
Integer

MatchQ[4,_Integer]
True

Head[4/3]
Rational

MatchQ[4/3,_Integer]
False
```

Putting these together, n_Plus means an expression which is labelled n and has the head Plus. Continuing with our example, x^2 matches "x to the power of an integer"

```
MatchQ[x^2, x^_Integer]
True

MatchQ[x^2, x^_Real]
False
```

In the same way x^2 matches "something or an expression to the power of 2"

```
MatchQ[x^2, _^2]
True

MatchQ[x^2, _^5]
False
```

Finally, x^2 matches "something to the power of something"

```
MatchQ[x^2,_^_]
True
```

One can put a condition on a pattern, i.e., we can test whether an expression satisfies a certain condition. Here is an example:

```
MatchQ[5, _Integer?(# > 3 &)]
True

MatchQ[2, _Integer?(# > 3 &)]
False
```

The pattern _Integer?(# > 3 &) stands for an expression which has Integer as its head, i.e., an integer, and is greater than three.

Here are some more examples:

```
MatchQ[x^2, _^_?OddQ]
False
```

```
MatchQ[x^2, _^_?EvenQ]
True
```

Here is how all these concepts help. One can single out an expression with a specific pattern and, once this is done, change the expression. It is all about accessing and then manipulating!

■ Problem 10.1

Show that if $a^3 = 0$ then $\frac{1}{2}a^2 + a + 1$ has the inverse $\frac{1}{2}a^2 - a + 1$. For example, a could be an $n \times n$ matrix such that $a^3 = 0$.

⟹ SOLUTION.

```
Expand[( 1/2 a^2 + a + 1) (1/2 a^2 - a + 1)]
1 + a^4/4
```

Below we replace any expression of the form a^_, where the exponent _ is an integer larger than 2, by 0. Namely, a^_?(# > 2 &) -> 0

```
Expand[( 1/2 a^2 + a + 1) (1/2 a^2 - a + 1)] /. a^_?(# > 2 &) -> 0
1
```

Here are some examples:

```
MatchQ[{a,b},{_,_}]
True
```

```
MatchQ[{a,b},{x_,y_}]
True
```

Study the following two examples carefully!

```
{{x1, x2}, {y1, y2}} /. {x_, y_} -> y
{y1, y2}
```

```
{{x1, x2}, {y1, y2}, {z1, z2}} /. {x_, y_} -> y
{x2, y2, z2}
```

In the first code x_ is matched by {x1,x2} and y_ by {y1,y2} and this is the reason the answer is {y1,y2}. However, in the second code, as there are three elements in the list, the only way *Mathematica* can match the pattern {x_, y_} to the list is to assign each of {x1, x2}, {y1, y2} and {z1, z2} to {x_, y_}. Problem 10.9 will demonstrate this situation and will show that if one is not careful with the pattern matching, things could go very wrong.

In the first code, if we need {x_, y_} to match each of the lists inside, we can do this by describing the pattern more precisely.

```
Head[x1]
Symbol

{{x1, x2}, {y1, y2}} /. {x_Symbol, y_} -> y
{x2, y2}
```

▬▬ Problem 10.2

Explain why

```
{1.1, 2, 3/2} /. n_ -> p[n]
p[{1.1, 2, 3/2}]
```

and change the code so that the result will be

```
{p[1.1], p[2], p[3/2]}
```

⟹ SOLUTION.

In the code {1.1, 2, 3/2} /. n_ -> p[n], the blank labelled by n matches the list {1.1, 2, 3/2}. Since the symbol p in the code is not listable, the result will be just p[1.1, 2, 3/2]. In order to lead *Mathematica* to match the blank with the numbers inside the list, we make sure to define the type of the blank as a number as follows:

```
{1.1, 2, 3/2} /. n_?NumberQ -> p[n]
{p[1.1], p[2], p[3/2]}
```

This will be used in Problem 15.3 where we change these numbers into points in the xy-plane by the rule:

```
{1.1, 2, 3/2} /. n_?NumberQ -> Point[{n, 0}]
{Point[{1.1, 0}], Point[{2, 0}], Point[{3/2, 0}]}
```

▬▬ Problem 10.3

Show that for any integer n greater than 1, the polynomial $n^4 + 64$ can be written as the product of four distinct positive integers greater than 1.

⟹ SOLUTION.

First, using Factor we decompose the polynomial into products of other polynomials.

```
Factor[n^12 + 64]
```

```
(2 - 2 n + n^2) (2 + 2 n + n^2) (4 - 4 n + 2 n^2 - 2 n^3 + n^4)
   (4 + 4 n + 2 n^2 + 2 n^3 + n^4)
```

This shows that $n^4 + 61$ can be written as a product of 4 numbers. Next we need to show that these numbers are distinct. We use Solve to check that if these polynomials represent the same number, then n has to be less than or equal to 1.

```
Solve[
  2 - 2 n + n^2 == 4 - 4 n + 2 n^2 - 2 n^3 + n^4, n, Integers]
```

```
{{n -> 1}}
```

Either we have to repeat this line with all the combination of these polynomials possible, or we can use pattern matching to approach it more efficiently.

```
(2 - 2 n + n^2) (2 + 2 n + n^2) (4 - 4 n + 2 n^2 - 2 n^3 + n^4)
   (4 + 4 n + 2 n^2 + 2 n^3 + n^4) /. a_*b_*c_*d_ -> {a, b, c, d}
```

```
{2 - 2 n + n^2, 2 + 2 n + n^2,
 4 - 4 n + 2 n^2 - 2 n^3 + n^4, 4 + 4 n + 2 n^2 + 2 n^3 + n^4}
```

```
l = Subsets[%, {2}]
```

```
{{2 - 2 n + n^2, 2 + 2 n + n^2}, {2 - 2 n + n^2,
  4 - 4 n + 2 n^2 - 2 n^3 + n^4}, {2 - 2 n + n^2,
  4 + 4 n + 2 n^2 + 2 n^3 + n^4}, {2 + 2 n + n^2,
  4 - 4 n + 2 n^2 - 2 n^3 + n^4}, {2 + 2 n + n^2,
  4 + 4 n + 2 n^2 + 2 n^3 + n^4}, {4 - 4 n + 2 n^2 - 2 n^3 + n^4,
  4 + 4 n + 2 n^2 + 2 n^3 + n^4}}
```

```
Equal @@@ l
```

```
{2 - 2 n + n^2 == 2 + 2 n + n^2,
 2 - 2 n + n^2 == 4 - 4 n + 2 n^2 - 2 n^3 + n^4,
 2 - 2 n + n^2 == 4 + 4 n + 2 n^2 + 2 n^3 + n^4,
 2 + 2 n + n^2 == 4 - 4 n + 2 n^2 - 2 n^3 + n^4,
 2 + 2 n + n^2 == 4 + 4 n + 2 n^2 + 2 n^3 + n^4,
 4 - 4 n + 2 n^2 - 2 n^3 + n^4 == 4 + 4 n + 2 n^2 + 2 n^3 + n^4}
```

```
Solve[#, n, Integers] & /@ %
```

```
{{{n -> 0}}, {{n -> 1}}, {}, {}, {{n -> -1}}, {{n -> 0}}}
```

This shows that the only cases where the polynomials agree are for $n = -1, 0$ or $n - 1$.

Exercise 10.1

Explain the results of the following codes.

```
{{a, b}, {c, d}} /. {x_, y_} -> x^ y
{a^c, b^d}

{{a, b}, {c, d}, {e, f}} /. {x_, y_} -> x^y
{a^b, c^d, e^f}

{{a, b}, {c, d}} /. {x_, y_} -> x y
{ac, bd}

{{a, b}, {c, d}, {e, f}} /. {x_, y_} -> x y
{a b, c d, e f}
```

The following Exercise asks us to provide an answer to Problem 2.8 using the power of pattern matching.

Exercise 10.2

The following formula generates many instance of the sum of three fourth powers being a fourth power. Check this with *Mathematica*.

$$(85v^2 + 484v - 313)^4 + (68v^2 - 586v + 10)^4 + (2u)^4 = (357v^2 - 204v + 363)^4,$$

where

$$u^2 = 22030 + 28849v - 56158v^2 + 36941v^3 - 31790v^4.$$

Exercise 10.3

Using pattern matching, find the smallest n such that the expression $x^{n-4} + 4n$ can be written as a product of two polynomials (with integer coefficients).

When plotting a graph, *Mathematica* produces a list of sample points and then connects these points together. In the following example, we access these sample points and manipulate them to create interesting graphs.

```
px3 = Plot3D[Sin[x^2 + y^2], {x, -Pi, Pi}, {y, -Pi, Pi},
    PlotPoints -> 50]
```

Here is how *Mathematica* stores data regarding this graph.

```
FullForm[px3]

Graphics3D[GraphicsComplex[List
List[List[-3.14159, -3.14159, 0.776852],
  List[-3.01336, -3.14159, 0.100243],
  List[-2.88514, -3.14159, -0.609907],
  List[-2.75691, -3.14159, -0.981742],
  List[-2.62868, -3.14159, -0.877966],
  List[-2.50045, -3.14159, -0.40218],
  .....
Rule[VertexNormals, List]]
```

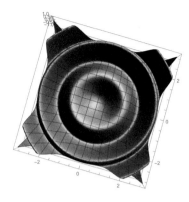

We are going to add some random value to the z coordinate of this graph.

```
xp3 = px3 /. {x_?NumberQ, y_?NumberQ, z_?NumberQ} :> {x, y,
    z + Random[]}
```

Exercise 10.4

Investigate what the following codes do.

```
px = Plot[Sin[x^2], {x, 0, 4}]
xp = px /. {x_?NumberQ, y_?NumberQ} :> {y, x};

Show[px, xp, PlotRange -> All, AspectRatio -> 1]
```

Here is a third approach to Problems 4.17 and 5.4.

▬▬ Problem 10.4

Write a function `squareFreeQ[n]` that returns `True` if the number n is a square free number, and `False` otherwise.

⟹ SOLUTION.

```
t=FactorInteger[234090]
{{2,1},{3,4},{5,1},{17,2}}
```

We are after those numbers such that, when decomposed into powers of primes, say, $\{\{p_1, k_1\}, \{p_2, k_2\}, \cdots, \{p_t, k_t\}\}$, all k_i are 1.

The pattern `{_,_?(#>1&)}` describes those lists with the second element (a number) bigger than 1.

```
MatchQ[{3,4},{_,_?(#>1&)}]
True
```

This is an appropriate time to introduce `Cases`.

```
? Cases
```

Cases[{e1, e2, ... }, pattern] gives a list of the ei that match the pattern.

```
Cases[{6,test,20,5.3,35,5/3},_Integer]
{6,20,35}
```

We use `Cases` to get all the pairs with k_i greater than 1. If this list is not empty then the number is not square free.

```
Cases[{{2,1},{3,4},{5,1},{17,2}},{_,_?(#>1&)}]
{{3,4},{17,2}}
```

```
Cases[{{2,1},{3,4},{5,1},{17,2}},{_,_?(#>1&)}] == {}
False
```

We are ready to put all of this together and write a function for finding square free numbers.

```
squareFree3[n_]:=Cases[FactorInteger[n],{_,_?(#>1&)}] == {}
```

```
squareFree3[234090]
False
```

```
squareFree3[3*5*13*17]
True
```

As with `Select` (see Problem 4.17), with `Cases` it is also possible to get only the first n expressions that satisfy a given pattern.

Cases[expr,pattern->rhs,levelspec] gives the values of rhs that match the pattern.

Cases[expr,pattern,levelspec,n] gives the first n parts in expr that match the pattern. >>

As in our problem, we only need to find one case of k_i being greater than 1, so it is enough to use Cases and single out just one of these items (again compare this with Problem 4.17).

```
squareFree3[n_]:=Cases[FactorInteger[n],{_,_?(#>1&)},1,1] == {}
```

```
squareFree3[234090]
False
```

```
squareFree3[3*5*13*17]
True
```

So far we have been dealing with one expression. What if, instead of one expression, we have to deal with a bunch of them?

```
MatchQ[{x^2},{_}]
True
```

```
MatchQ[{x^2, x^3, x^5}, {_}]
False
```

```
MatchQ[{x^2, x^3, x^5}, {__}]
True
```

As one can see from the above example, __ stands for a sequence of data whereas _ stands for just one expression. In fact __ is for a sequence of nonempty expressions, and ___ is for a sequence of empty or more data. The following examples show this clearly.

```
MatchQ[{},{_}]
False
```

```
MatchQ[{},{__}]
False
```

```
MatchQ[{},{___}]
True
```

Here is one more example which illustrates the difference between __ and ___:

```
MatchQ[{3,5,2,2,stuff,7},{__,3,___}]
False
```

```
MatchQ[{3,5,2,2,stuff,7},{___,3,___}]
True
```

```
MatchQ[{3,5,2,2,7,us},{___,2,2,___}]
True
```

```
{one, two, three, four} /. {x_, y___, z_} -> {z, y^z, y/z}
{four, two^three^four, (three two)/four}
```

Exercise 10.5

Explain what the following code does.

```
datasample = Table[Random[Integer, {1, 10}], {1000}];
frequence[data_List, n_] :=
  Apply[Plus, data /. n -> "a" /. x_Integer -> 0 /. "a" -> 1];
Table[frequence[datasample, n], {n, 1, 10}]
```

■ Problem 10.5

In the first 50000 prime numbers, find those that have 2009 embedded in them
(e.g., 420097 is prime and 2009 is sitting in it).

⟹ SOLUTION.

Here are two quite similar solutions: the second solution uses a technique
similar to Exercise 10.5.

```
(* First solution *)

Select[IntegerDigits /@ Prime /@ Range[1, 50000],
  MatchQ[#, {m___, 2, 0, 0, 9, n___}] &]

{{3, 2, 0, 0, 9}, {5, 2, 0, 0, 9}, {8, 2, 0, 0, 9}, {9, 2, 0, 0,
  9}, {1, 2, 0, 0, 9, 1}, {1, 2, 0, 0, 9, 7}, {1, 7, 2, 0, 0,9},
  {1, 8, 2, 0, 0, 9}, {2, 0, 0, 9, 0, 3}, {2, 0, 0, 9, 0, 9},
  {2, 0, 0, 9, 2, 7}, {2, 0, 0, 9, 2, 9}, {2, 0, 0, 9, 7, 1},
  {2, 0, 0, 9, 8, 3}, {2, 0, 0, 9, 8, 7}, {2, 0, 0, 9, 8, 9},
  {2, 4, 2, 0, 0, 9}, {2, 7, 2, 0, 0, 9}, {3, 0, 2, 0, 0, 9},
  {3, 2, 2, 0, 0, 9}, {3, 3, 2, 0, 0, 9}, {4, 2, 0, 0, 9, 7},
  {4, 4, 2, 0, 0, 9}, {4, 5, 2, 0, 0, 9}, {5, 1, 2, 0, 0, 9},
  {5, 3, 2, 0, 0, 9}}

(* second solution *)
t =
Select[IntegerDigits /@
    Prime /@ Range[1, 50000] /. {m___, 2, 0, 0, 9, n___} ->
    {X, m, 2, 0, 0, 9, n}, MemberQ[#, X] &]

{{X, 3, 2, 0, 0, 9}, {X, 5, 2, 0, 0, 9}, {X, 8, 2, 0, 0,
  9}, {X, 9, 2, 0, 0, 9}, {X, 1, 2, 0, 0, 9, 1},
  {X, 1, 2, 0, 0, 9, 7}, {X, 1, 7, 2, 0, 0, 9},
  {X, 1, 8, 2, 0, 0, 9}, {X, 2, 0, 0, 9, 0, 3},
  {X, 2, 0, 0, 9, 0, 9}, {X, 2, 0, 0, 9, 2, 7},
  {X, 2, 0, 0, 9, 2, 9}, {X, 2, 0, 0, 9, 7, 1},
  {X, 2, 0, 0, 9, 8, 3}, {X, 2, 0, 0, 9, 8, 7},
  {X, 2, 0, 0, 9, 8, 9}, {X, 2, 4, 2, 0, 0, 9},
  {X, 2, 7, 2, 0, 0, 9}, {X, 3, 0, 2, 0, 0, 9},
  {X, 3, 2, 2, 0, 0, 9}, {X, 3, 3, 2, 0, 0, 9},
  {X, 4, 2, 0, 0, 9, 7}, {X, 4, 4, 2, 0, 0, 9},
  {X, 4, 5, 2, 0, 0, 9}, {X, 5, 1, 2, 0, 0, 9},
  {X, 5, 3, 2, 0, 0, 9}}
```

```
FromDigits /@ Rest /@ t
{32009, 52009, 82009, 92009, 120091, 120097, 172009,
182009, 200903, 200909, 200927, 200929, 200971, 200983, 200987,
200989, 242009, 272009, 302009, 322009, 332009, 420097, 442009,
452009, 512009, 532009}
```

Exercise 10.6

Observe that $\mathtt{Range[10]/.\{x\_,y\_\_\_\}\rightarrow y/x}$ amounts to 10!

We are ready to write a little game.

▬▬ Problem 10.6

Write a game as follows. A player is randomly dealt 7 cards between 1 and 9. He is able to discard any two cards between 4 and 9 that are the same. Then the sum of the cards that remain in the hand is what a player scores. The player with the lowest score wins.

$\Longrightarrow$ SOLUTION.

First, we generate a list containing 7 random numbers between 1 and 10.

```
s=Table[RandomInteger[{1, 9}], {7}]
{2,5,2,3,4,7,4}
```

Now we shall write a code to discard any two numbers which are the same. The trick we use here is, we look inside the list and recognise the same numbers (which have the same pattern), mark them and with a rule send the list to a new list containing all the elements except the similar ones. Here is the code:

```
s /. {m___, x_, y___, x_, n___} -> {m, y, n}
{5,3,4,7,4}
```

Here *Mathematica* looks for similar expressions x_ and x_. To the left of x_ is m___ which means any sequence of empty or more expressions. Similarly in between x's we place y___. That is, in between similar numbers could be nothing (i.e., the similar numbers could be next to each other) or a bunch of other expressions. Finally, to the right of the second x_ is n___. In the example, our original list is {2,5,2,3,4,7,4}. *Mathematica* recognises that there are two 2's in the list, so will assign them to x_. To the left of the first 2 there is no data, thus m___ would return an empty value, y___ would be 5 and n___ would be the whole sequence of 3, 4, 7, 4 to the right of the second 2. Thus the rule{m___,x_,y___,x_,n___}->{m,y,n} will discard the x's and give us {5,3,4,7,4}.

Still there are two 4's in the list but, as we have seen in Chapter 9, /.->
would go through the list only once. Thus if we run the same code again, this
time with our new list, we will get rid of the double 4's.

```
{5, 3, 4, 7, 4} /. {m___, x_, y___, x_, n___} -> {m, y, n}
{5, 3, 7}
```

Remember that //.-> was designed exactly for this job.

```
s //. {m___, x_, y___, x_, n___} -> {m, y, n}
{5,3,7}
```

But in the game we are allowed to discard the cards between 4 and 10. Thus
we shall include a little test to find numbers larger than 3 which are the same.
Here is the enhanced code:

```
s //. {m___, x_?(3 < # < 10 &), y___, x_, n___} -> {m, y, n}
{2,5,2,3,7}
```

Notice that it is enough to perform the test for one of the x_'s.
Just to make sure we understood this, let's try the code for

```
s = {7, 6, 2, 7, 1, 2, 1, 6, 7};
```

```
s //. {m___, x_?(3 < # < 10 &), y___, x_, n___} -> {m, y, n}
{2, 1, 2, 1, 7}
```

It remains to sum the numbers in the list. For example

```
Plus @@ %
13
```

L_____

_____ Problem 10.7

A *Kaprekar* number is a number such that its square can be split into two
(positive) integer parts whose sum is equal to the original number (e.g. 45,
since $45^2 = 2025$, and $20 + 25 = 45$). Find all 5-digit Kaprekar numbers.

⟹ SOLUTION.
   Let us start with the number 45. We shall look at $45^2 = 2025$, then consider
$2 + 025$, $20 + 25$ and $202 + 5$. Thus we first write a program to create a list
of all these arrangements. For this we need the command ReplaceList, which
is similar in spirit to the commands Replace and ReplaceRepeated which we
saw in Chapter 9.

```
?ReplaceList
```

ReplaceList[expr,rules] attempts to transform the entire
expression expr by applying a rule or list of rules in all
possible ways, and returns a list of the results obtained

```
ReplaceList[{2, 0, 2, 5}, {x__, y__} -> {{x}, {y}}]
{{{2}, {0, 2, 5}}, {{2, 0}, {2, 5}}, {{2, 0, 2}, {5}}}
```

We are almost there (except there are more curly brackets here than we would
like). If we Map the command FromDigits, which puts these digits together and
returns a number, we get all the arrangements we are looking for. We need to
push FromDigits inside the second list, that is, it needs to be applied in the
second level of the list we have. For this we use Map[f,list,{2}] where the 2
tells *Mathematica* to apply the map f to the second level in the list.

```
Map[FromDigits,
  ReplaceList[{2, 0, 2, 5}, {x__, y__} -> {{x}, {y}}], {2}]
{{2, 25}, {20, 25}, {202, 5}}
```

The rest is (almost) clear. If we replace the inside lists with Plus (i.e., changing
heads) we get the sum of all the numbers and this is exactly what we are after.
Again, we want to apply Plus to the second layer (or level) of the list. For
this, as in Map, we need to use Apply[f,expr,2] where the 2 refers to the
second level. Remember the shorthand for Apply (to the first layer) is @@ and
for applying to the second layer is @@@, so we have

```
Plus @@@ (Map[FromDigits, ReplaceList[{2, 0, 2, 5},
{x__, y__} -> {{x}, {y}}], {2}])
{27, 45, 207}
```

This already shows 45 is a Kaprekar number as we see the number 45 in
the list above. We are ready to write the whole list:

```
Select[Range[10000,99999],
  MemberQ[Plus @@@ (Map[FromDigits,
    ReplaceList[
      IntegerDigits[#^2], {x__, y__} -> {{x}, {y}}], {2}]), #] &]
```

{10000, 17344, 22222, 38962, 77778, 82656, 95121, 99999}

Recall that the Kaprekar numbers are those whose square can be written
as a concatenation of *positive* integer parts whose sum is equal to the original
number. This excludes 10, 100 and 10000 from this list as $10^2 = 100$ and
$10 = 10 + 0$ but 0 is not positive. The reader is encouraged to modify the code
to remove the cases of this nature.

Exercise 10.7

Find all the words in the *Mathematica* dictionary which end with "rat".

_____ Problem 10.8

Find all the words in the *Mathematica* dictionary which contain the letters
"c","u", "t" and "e".

⟹ SOLUTION.
    We need the function Characters:

```
?Characters
Characters["string"] gives a list of the characters in a string.

Characters["Freshwater"]
{"F", "r", "e", "s", "h", "w", "a", "t", "e", "r"}


cute = Select[DictionaryLookup[],
    MatchQ[Characters[#], {___, "c", ___, "u", ___, "t", ___,
        "e", ___}] &];

Short[cute]

{accentuate,accentuated, accentuates, <<700>>, woodcutter,
woodcutters}

Length[cute]
705
```

One needs to be careful when using pattern matching, as the following
problem demonstrates.

_____ Problem 10.9

Write the Syracuse function, which is defined as follows: for any odd positive
integer n, $3n+1$ is even, so one can write $3n+1 = 2^k n'$ where k is the highest
power of 2 that divides $3n+1$. Define $f(n) = n'$. Show that for any odd positive
integer m, applying f repeatedly, we arrive at 1.

⟹ SOLUTION.
    It is not difficult to see that this is in fact a different version of the Col-
latz function (see Problems 6.6 and 11.2). Here is one way to write the function. Recall that FactorInteger gives the decomposition of a number into its
prime factors, i.e., if $n = 2^{k_1} 3^{k_2} \cdots p_i^{k_i}$, then using FactorInteger we get
$\{\{2, k_1\}, \{3, k_2\}, \cdots, \{p_i, k_i\}\}$. So, if we drop $\{2, k_1\}$ from the list and multiply
the rest together, the result is the n' that we are looking for. Here is one way
to do so:

```
FactorInteger[2^5*3*5*7]
{{2, 5}, {3, 1}, {5, 1}, {7, 1}}

Rest[FactorInteger[2^5*3*5*7]]
 {{3, 1}, {5, 1}, {7, 1}}

Rest[FactorInteger[2^5*3*5*7]] /. {x_, y_} -> x^y
{3, 5, 7}

Times @@ (Rest[FactorInteger[2^5*3*5*7]] /. {x_, y_} -> x^y)
105

f[n_?OddQ] := Times @@ (Rest[FactorInteger[3n+1]] /.
{x_, y_} -> x^y)
```

We test the function for smaller values.

```
f[7]
11

7*3 + 1
22

FactorInteger[22]
{{2, 1}, {11, 1}}
```

Now that f is defined, we are going to apply f repeatedly to a given number until we reach one. For this we will use FixedPointList, introduced in Problem 8.14.

```
FixedPointList[f, 133]
{133, 25, 19, 29, 11, 17, 13, 5, 1, 1}
```

So far, so good. Let us start with 123.

```
FixedPointList[f, 123]

During evaluation of In[196]:= General::ovfl: Overflow occurred
in computation. >>

During evaluation of In[196]:= FactorInteger::exact: Argument
Overflow[] in FactorInteger[Overflow[]] is not an exact number.

During evaluation of In[196]:= FactorInteger::argt:FactorInteger
called with 0 arguments; 1 or 2 arguments are expected. >>

Out[196]= {123, 7275957614183425903203125,
682121026329696178436279B, 5115907697472721338272095,
7673861546209082007408143, 1151079231931362301111221E,
172661884789704345166668323, 258992827184556517750024BE,
Overflow[], 1, 1}
```

What seems to be the problem? Let us look at the rule we have defined:

```
{{5, 1}, {37, 1}, {3, 2}} /. {x_, y_} -> x^y
{5, 37, 9}
```

This is what we wanted. Let us try another example:

```
{{5, 1}, {37, 1}} /. {x_, y_} -> x^y
{727595761418342590332031125, 1}
```

Something has gone horribly wrong here. What we wanted was {5, 37}. The problem is, in the last example, since the list itself contains two lists, then x_ takes {5,1} and y_ takes {37,1}. To help *Mathematica* to get out of this confusion, we can describe the pattern of x_ more precisely, namely, x_ stands for an integer and not a list.

```
{{5, 1}, {37, 1}} /. {x_Integer, y_} -> x^y
{5, 37}
```

This is the correct approach. We redefine the function f according to this new rule.

```
Clear[f]
```

```
f[n_?OddQ] := Times @@ (Rest[FactorInteger[3n+1]] /.
{x_Integer, y_} -> x^y)
```

```
FixedPointList[f, 123]
{123, 185, 139, 209, 157, 59, 89, 67, 101, 19, 29, 11, 17,
13, 5, 1, 1}
```

```
FixedPointList[f, 133]
{133, 25, 19, 29, 11, 17, 13, 5, 1, 1}
```

Finally, let us mention that there is another way to define the function using IntegerExponent:

```
?IntegerExponent
IntegerExponent[n,b] gives the highest power of b that
divides n. >>
```

```
f[n_] := Quotient[3 n + 1, 2^IntegerExponent[3 n + 1, 2]]
```

```
f[2^5*3*5*7]
105
```

```
FixedPointList[f, 123]
{123, 185, 139, 209, 157, 59, 89, 67, 101, 19, 29, 11, 17,
13, 5, 1, 1}
```

We briefly mentioned how to define a function denoted by ⊕ in Chapter 3 (see also Problem 8.17). Here we look at this feature systematically.

_____ Problem 10.10

Define a binary operation on real numbers as follows:

$$\oplus : \mathbb{R} \times \mathbb{R} \longrightarrow \mathbb{R}$$
$$(x, y) \longmapsto x \oplus y = x + y + xy.$$

Using *Mathematica*, show that this operation is associative, i.e.,

$$(x \oplus y) \oplus x = x \oplus (y \oplus z).$$

$\Longrightarrow$ SOLUTION.

Here is how we define the binary operation.

```
x_⊕y_:=x+y+x*y
```

The symbol $\oplus$ can be typed by pressing the keys "esc" c+ "esc".

The idea is, anywhere *Mathematica* detects a pattern of the form x⊕y, it will change it to x+y+x*y. We check this in fact operates correctly.

```
x⊕0= x
x⊕-x= -x²
```

Now we are in a position to use this definition and show this operation is associative.

```
Simplify[(x⊕y)⊕z==x⊕(y⊕z)]
True
```

Next we look at the repeated patterns. Study the following examples:

```
MatchQ[{1, 2}, {__?NumberQ}]
True

MatchQ[{{1, 2}, {3, 4}}, {__?NumberQ}]
False

MatchQ[{{1, 2}, {3, 4}}, {{__?NumberQ} ..}]
True

?..
p.. or Repeated[p] is a pattern object that represents a
sequence of one or more expressions, each matching p.
```

Here are further examples showing how to produce repeated patterns and how to describe them using ..

```
Range[Range[3]]
{{1}, {1, 2}, {1, 2, 3}}

MatchQ[Range[Range[3]], {{__?NumberQ} ..}]
True

Nest[Range, 3, 1]
{1, 2, 3}

Nest[Range, 3, 2]
{{1}, {1, 2}, {1, 2, 3}}

Nest[Range, 3, 3]
{{{1}}, {{1}, {1, 2}}, {{1}, {1, 2}, {1, 2, 3}}}

MatchQ[Nest[Range, 3, 3], {{__?NumberQ} ..}]
False

MatchQ[Nest[Range, 3, 3], {{{__?NumberQ} ..} ..}]
True
```

As the above examples show, {{__?NumberQ} ..} stands for a pattern of list of lists, each containing numbers. Repeating further {{{__?NumberQ} ..} ..} stands for a list with three layers each containing numbers. In Problem 11.4 we will use these patterns to plot specific types of data.

In this chapter we will talk about functions with multiple definitions. Also we will see how a function can contain more than one line, that is, have a block of codes with its own local variables.

In Chapter 3, we defined functions in Wolfram *Mathematica*® which consisted of one line. In Section 6.3, we defined functions which came with conditions, using If, Which and Piecewise.

In this chapter we will talk about the ability of *Mathematica* to handle a function with multiple definitions. Also we will see how a function can contain more than one line, i.e., contain a block of codes (a sort of mini-program or a procedure).

Recall the very first function that we defined in Section 3.1.

```
f[n_] := n^2 + 4
```

```
f[13]
173
```

In light of the previous chapter, one can see exactly what this code means. One can send any expression with any pattern into f[n_]. The expression is labelled n. Now we can easily restrict the sort of data we want to send into a function, by simply describing the sort of pattern we desire. For example, if in the above function we would like the function only to take on positive integers, then

```
f[n_Integer?Positive] := n^2 + 4
```

```
f[4]
20
```

```
f[-2]
f[-2]
```

© Springer International Publishing Switzerland 2015

R. Hazrat, *Mathematica*®*: A Problem-Centered Approach*, Springer Undergraduate Mathematics Series, DOI 10.1007/978-3-319-27585-7_11

Here are some more examples:

```
g[n_Integer?(0 < # < 5 &)] := Sqrt[5 - n]

g[1]
2

g[6]
g[6]

g[2.6]
g[2.6]

e[p__?(PolynomialQ[#, x] &)] := Expand[p, x]

e /@ {4, (1 + x)^2, (1 + y)^2, (Sin[x] + Cos[x]^2)}
{4, 1 + 2 x + x^2, (1 + y)^2, e[Cos[x]^2 + Sin[x]]}
```

One can even be carried away with this ability. Here is a function that gives the prime factors of a number which consists only of odd digits (e.g. 3715).

```
myfunc[n_Integer?(Select[IntegerDigits[#], EvenQ, 1] == {}&)]:=
Map[First, FactorInteger[n]]

myfunc[3715]
{5, 743}

myfunc[593183]
myfunc[593183]
```

One of the great features of *Mathematica* is that one can define a function with multiple definitions. Here is a harmless example

```
oddeven[(n_?EvenQ)?Positive] := Print[n, " even and positive"]
oddeven[(n_?EvenQ)?Negative] := Print[n, " even and negative"]
oddeven[(n_?OddQ)?Positive] := Print[n, " odd and positive"]
oddeven[(n_?OddQ)?Negative] := Print[n, " odd and negative"]

Map[oddeven, {-2, 5, -3, -4}];
-2  even and negative
5 odd and positive
-3   odd and negative
4   even and positive
```

Here we have the function oddeven with four definitions. An integer falls into one of the cases above, and *Mathematica* has no problem going through all the definitions of the function and applying the appropriate one to the given number. If one asks for the definition of oddeven, one can see *Mathematica* has all four definitions in memory, in the same order that one has defined the function.

```
?oddeven

Global'oddeven

oddeven[(n_?EvenQ)?Positive] := Print[n,   even and positive]

oddeven[(n_?EvenQ)?Negative] := Print[n,   even and negative]

oddeven[(n_?OddQ)?Positive] := Print[n,   odd and positive]

oddeven[(n_?OddQ)?Negative] := Print[n,   odd and negative]
```

──── Problem 11.1

Define the function

$$f(x) = \begin{cases} \sqrt{x} & \text{if } x \geq 0 \\ \sqrt{-x} & \text{if } x < 0 \end{cases}$$

and plot the graph of the function for $-1 \leq x \leq 1$.

⟹ SOLUTION.

One can define $f(x)$ in *Mathematica* as a function with two definitions as follows:

```
f[x_?NonNegative] := Sqrt[x]
f[x_?Negative]    := Sqrt[-x]
Plot[f[x],{x,-1,1}]
```

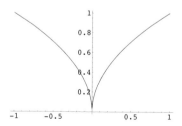

Figure 11.1 A function with multiple definitions

See Chapter 14 for more on graphics.

As you might have noticed so far, there has been no confusion regarding the multiple definitions of a function. That is, the data that is sent to the function satisfied only one of the patterns in the definition of the function. In oddeven, a number could only be one of the cases of positive/negative and odd/even and

in the previous example a number is either positive or negative. But suppose
we define a function as follows:

```
f[x_] := x
f[x_Integer] := x!
```

Now one might ask what f[4] would be. There are two definitions for f
and 4 can match both patterns, namely x_ or x_Integer.

```
f[4]
24

f[5]
120

f[2.3]
2.3

f[test]
test
```

Thus for any integer the definition which is the factorial of a number is
performed and for other data the other definition is used (obviously). If we find
out in what order *Mathematica* saves the definitions of functions, we can justify
this action.

```
?f

Global'f

f[x_Integer] := x
f[x_] := x
```

Thus, in principle, *Mathematica* stores the definitions from the one with
more precise pattern matching (here the one with x_Integer). If she cannot
decide which definition has the more precise pattern matching, then she stores
definitions in the order in which they have been entered into the system.

▬▬ Problem 11.2

Define the Collatz function as follows:

$$f(x) = \begin{cases} x/2 & \text{if } x \text{ is even} \\ 3x + 1 & \text{if } x \text{ is odd.} \end{cases}$$

Find out how many times one needs to apply f to numbers 1 to 200 to get 1.

⟹ Solution.
We have defined this function in Problem 6.6 using If. Another version of
this problem was discussed in Problem 10.9.
Here we simply use a multi-definition function to define the Collatz function.

```
f[x_Integer?EvenQ] := x/2

f[x_Integer] := 3x + 1
```

As we mentioned above, the natural question which would be raised here is what is f[2], as both definitions of f can accept this number. If cases like this happen, *Mathematica* chooses the definition which has the more precise description for the given expression. In the above definitions, although 2 is an integer and so fits in _Integer, but _Integer?EvenQ is a more accurate description for 2 and thus *Mathematica* chooses f[x_Integer?EvenQ] := x/2 for evaluating f[2]. If *Mathematica* cannot decide in this manner, she would simply use the function which was defined first.

One can write the following one-liner for the rest of the code.

```
l=Length /@ ( NestWhileList[f, #, #  !== 1 &] & /@ Range[200])
```

```
{1, 2, 8, 3, 6, 9, 17, 4, 20, 7, 15, 10, 10, 18, 18, 5, 13, 21,
21, 8, 8, 16,  16, 11, 24, 11, 112, 19, 19, 19, 107, 6, 27, 14,
14, 22, 22, 22, 35,9, 110,   9, 30, 17, 17, 17, 105, 12, 25, 25,
25, 12, 12, 113, 113, 20, 33, 20, 33, 20,  20, 108, 108, 7, 28,
28, 28, 15, 15, 15, 103, 23, 116, 23, 15, 23, 23, 36,  36, 10,
23, 111, 111, 10, 10, 31, 31, 18, 31, 18, 93, 18, 18, 106, 106,
13, 119, 26, 26, 26, 26, 26, 88,13, 39, 13, 101, 114, 114, 114,
70, 21, 13, 34,  34, 21, 21, 34, 34, 21, 96, 21, 47, 109, 109,
109, 47, 8, 122, 29, 29, 29,  29, 29, 42, 16, 91, 16, 42, 16,
16, 104, 104, 24, 117, 117, 117, 24, 24, 16,  16, 24, 37, 24,
86, 37, 37,37, 55, 11, 99, 24,24, 112, 112, 112, 68,11, 50, 11,
125, 32, 32, 32, 81, 19,32, 32, 32, 19, 19, 94, 94, 19, 45, 19,
45,  107, 107, 107, 45, 14, 120, 120, 120, 27, 27, 27, 120, 27}
```

```
Max[%]
```

125

```
ListPlot[l, Filling -> Axis]
```

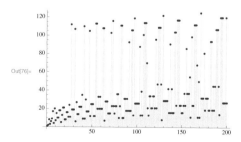

Figure 11.2 The Collatz function

In this example, *Mathematica* stores the definitions of the function in the

same order that we entered it, as there is no preference in the patterns that
have been defined.

```
?f

Global'f

f[x_Integer?EvenQ]:=x/2

f[x_Integer]:=3 x+1
```

▬▬ Problem 11.3

Define a function $f(x)$ in *Mathematica* which satisfies

$$f(xy) = f(x) + f(y)$$
$$f(x^n) = nf(x)$$
$$f(n) = 0$$

where n is an integer and show that $f\left(\prod_{i=1}^{20} i(x_i)^i\right) = \sum_{i=1}^{20} if(x_i)$.

$\Longrightarrow$ SOLUTION.

We first translate the function into *Mathematica*.

```
f[x_ y_] := f[x] + f[y]
f[x_^n_Integer] := n f[x]
f[n_Integer] := 0

f[Product[i(x_i)^i, {i, 1, 20}]]
f[x_1] + 2 f[x_2] + 3 f[x_3] + 4 f[x_4] + 5 f[x_5]
+ 6 f[x_6] + 7 f[x_7] + 8 f[x_8] + 9 f[x_9] + 10
f[x_10] + 11 f[x_11] + 12 f[x_12] + 13 f[x_13] + 14
f[x_14] + 15 f[x_15] + 16 f[x_16] + 17 f[x_17] + 18
f[x_18] + 19 f[x_19] + 20 f[x_20]

∑_i^20 i f[x_i]== f[Product[i (x_i)^i, {i, 1, 20}]]
True
```

—— Problem 11.4

Define the function dplot, which accepts dates and data, and plot the graph using DateListPlot. The function should accepts lists of different types as follows:

1. {{{date1},d1},{{date2},d2},...,{{daten},dn}}

2. {{{{date1},d11},{{date2},d12},...,{{daten},d1n}},
   {{{date1},d21},{{date22},d2},...,{{daten},d2n}}
   ...
   {{{date1},dk1},{{date2},dk2},...,{{daten},dkn}}}

3. {{{d11,d12,...,d1n},{d21,d22,...,d2n},...,{dk1,dk2,...,dkn}},
   {date1,date2,...,daten}}

⟹ SOLUTION.

We start with an example of a data set.

```
data={{{{1998, 1}, 1}, {{1999, 2}, 2}, {{2000, 2}, -1}},
      {{{1998, 1}, 2}, {{1999, 2}, -1}, {{2000, 2}, 3}},
      {{{1998, 1}, -3}, {{1999, 2}, -4}, {{2000, 2}, 2}}}
```

```
DateListPlot[data]
```

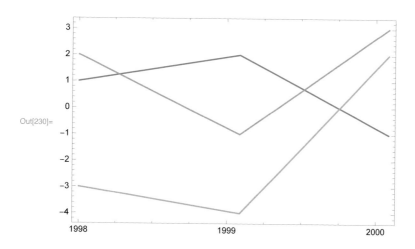

Out[230]=

Let us understand the pattern data for the definition of other functions.

```
MatchQ[{{1998, 1}, 1}, {_List, _}]
True

MatchQ[{{{1998, 1}, 1}, {{1999, 2}, 2}, {{2000, 2}, -1}},
    {{_List, _} ..}]
True

MatchQ[data, {{{_List, _} ..} ..}]
True

data[[1]]
{{{1998, 1}, 1}, {{1999, 2}, 2}, {{2000, 2}, -1}}

DateListPlot[data[[1]]]
```

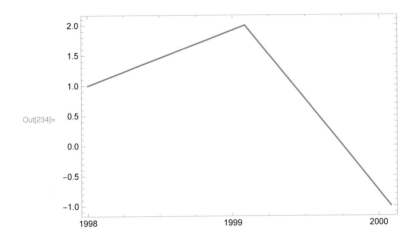

Out[234]=

Based on the examples above, here are the first two functions for dplot, handling different types of data sets.

```
dplot[da : {{{_List, _} ..} ..}] :=
DateListPlot[da, Joined -> True]

dplot[da : {{{_List, _} ..} ..}, n_Integer] :=
DateListPlot[da[[n]], Joined -> True]
```

Next we define the function dplot to handle data of the type (3) in the problem.

```
data = {{1, 2, 3}, {-2, 5, -1}, {1, 1, 1}}
```

```
MatchQ[data, {{__?NumericQ} ..}]
True
```

```
dates = {{1998}, {1999}, {2000}}
{{1998}, {1999}, {2000}}
```

We would like to feed the following type of data to `DateListPlot`.

```
{data, dates}
```

```
{{{1, 2, 3}, {-2, 5, -1}, {1, 1, 1}}, {{1998}, {1999}, {2000}}}
```

```
DateListPlot[{data, dates}]
```

```
During evaluation of In[247]:= DateListPlot::ldata:
{{{1,2,3},{-2,5,-1},{1,1,1}},{{1998},{1999},{2000}}} is not a
valid dataset or list of datasets. >>
```

For this reason, we need to change the format of the data given.

```
Length[dates] == Length[data[[1]]]
True
```

```
data[[All, 1]]
{1, -2, 1}
```

```
data[[All, 2]]
{2, 5, 1}
```

```
{dates, data[[All, #]]} & /@ Range[Dimensions[data][[2]]]
```

```
{{{{1998}, {1999}, {2000}}, {1, -2,
   1}}, {{{1998}, {1999}, {2000}}, {2, 5,
   1}}, {{{1998}, {1999}, {2000}}, {3, -1, 1}}}
```

We are almost there. We need to assign the years to each of the data. The command `Transpose` will do the trick, as the following shows.

```
{dates, {1, 2, 3}}
{{{1998}, {1999}, {2000}}, {1, 2, 3}}
```

```
Transpose[{dates, {1, 2, 3}}]
{{1998, 1}, {1999, 2}, {2000, 3}}
```

```
Transpose[{dates, data[[All, #]]}] & /@ Range[Dimensions[data][[2]]]
{{{{1998}, 1}, {{1999}, -2}, {{2000}, 1}}, {{{1998}, 2}, {{1999},
   5}, {{2000}, 1}}, {{{1998}, 3}, {{1999}, -1}, {{2000}, 1}}}
```

We now define the function.

```
dplot[data : {{__?NumericQ} ..}, dates_List] :=
dplot[Transpose[{dates, data[[All, #]]}] & /@
   Range[Dimensions[data][[2]]]]
```

Note that in the definition of the function above, we first change the pattern of the data to a suitable form and then call dplot, which has already been defined, which handles this suitable type of data.

```
dplot[data, dates]
```

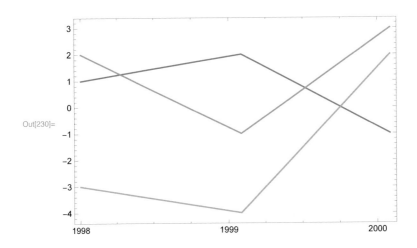

11.1 Functions with local variables

One of the approaches of procedural languages (like C) to programming is to break the program into "mini-programs" or *procedures* and then put them together to get the code we need. These procedures have their own variables called *local* variables, that is, variables which have been defined only inside the procedure. So far all the functions that we have defined consist of only one line. *Mathematica*'s functions can also be used as procedures, i.e., they can contain several lines of code and their own local variables. Let us look at a simple example. Recall Problem 8.8, which finds all the prime numbers less than n. Let us write this as a function 1Primes[n] to produce a list of all such primes.

```
1Primes[n_] := Module[{pset = {}, i = 1},
While[Prime[i] <= n,
pset = pset ∪ {Prime[i]};
i++];
pset]
```

```
1Primes[8]
{2, 3, 5, 7}
```

A function with several lines of code in *Mathematica* is wrapped by Module. The structure looks like Module[{local variables},body]. In the above example the variables pset and i are variables defined only inside the function 1Primes. Here we check that these are undefined outside the function:

```
pset
pset
```

```
i
i
```

There are two other ways in *Mathematica* to collect codes, define local variables and make a mini-program, namely using With and Block.

The command Block is mainly used to temporarily change settings of global system parameters. We have already seen one example in Section 1.1.

```
$MaxExtraPrecision
50.
```

```
N[Sin[(E + Pi)^100], 60]
```

During evaluation of In[18]:= N::meprec: Internal precision limit $MaxExtraPrecision=50. reached while evaluating Sin[(E+Pi)]^100]

0.799751375979095575644472881373733

```
Block[{$MaxExtraPrecision = 1000}, N[Sin[(E + Pi)^100], 60]]
```

0.79975137597909557564447288137325189976626145567341256760739

We will see another example of Block in Chapter 12, where we will change the recursion limit.

The command With is used to define a *constant* throughout the block of the code. One cannot change this constant inside the block.

```
With[{c = 5}, x = c + 10]; x
15
```

```
c
c
```

There are some fine differences between `Module`, `Block` and `With` and how they handle the local variables. See the *Mathematica* Document Center for more examples and explorations.

11.2 Functions with conditions

Consider the following code.

```
f[n_] := Sqrt[n] /; n >0

f[4]
2

f[-4]
f[-4]
```

Here /; is a shorthand for `If`. We have seen that we can restrict the pattern of the data we pass into a function. The *almost* equivalent ways to define the above function are

```
f1[n_?Positive]:=Sqrt[n]

f2[n_]:=If[n>0,Sqrt[n]]
```

The following shows what the difference between these functions is:

```
f[4]
2

f1[4]
2

f2[4]
2

f[-3]
f[-3]

f1[-3]
f1[-3]

f2[-3]
```

That is, the one which is defined by `If` would return `Null` if the argument does not satisfy the condition.

Sometimes using /; helps to make the code much more readable than using alternative methods.

Here is another version of the game in Problem 10.6.

___ Problem 11.5

Write a game as follows. A player is randomly dealt 7 cards between 1 and 9. He can discard any two cards with the sum 5. Then the sum of the cards that remain in the hand is what a player scores. The player with the lowest score wins.

$\Longrightarrow$ SOLUTION.

Let us first design a function that accepts a sequence of numbers and deletes any two numbers of whose sum is 5. Having an eye on the code of Problem 10.6 we proceed as follows:

```
aHandD[n___, y_, t___, z_, m___] := aHandD[n, t, m] /; y + z == 5

aHandD[2,3,5,4,1,3,7]
aHandD[5,3,7]
```

The function is defined as follows: it looks inside the list of arguments (here 2,3,5,4,1,3,7) and identifies the numbers whose sum is 5 (here, 1 and 4). Assign those to y and z. Now the function is defined as aHandD[n,t,m], that is, it drops y and z from the list. Thus the list, or cards, becomes 2,3,5,3,7. Now what is important here is that this is a recursive function. That is, once the numbers y and z are dropped, the function looks at the remaining arguments and tries to identify the next two numbers whose sum is 5. This is going to be repeated until there are no such numbers. This is called a recursive function, which is the theme of Chapter 12. The rest is easy, we want the sum of the remaining numbers. So we can simply change the head of this expression to Plus to get the sum of the cards.

```
Apply[Plus,%]
15
```

Now we produce a list of 7 random numbers and write a little function, call it aHand, to put all these lines together:

```
Table[RandomInteger[{1,9}],{7}]
{5,2,7,3,1,6,3}

Apply[Sequence,%]
Sequence[5,2,7,3,1,6,3]

aHandD[%]
aHandD[5,7,1,6,3]

aHand=Module[{},
aHandD[Apply[Sequence, Table[Random[Integer, {1,
9}], {7}]]];
  Print[Apply[Plus, %]]
  ]
```

We shall see a similar approach to a problem involving matrices in Problem 13.4.

Exercise 11.1

Consider the following function and write the function using Which.

```
t[x_, y_, z_] /; x == y := x

t[x_, y_, z_] := z
```

11.3 Functions with default values

In Chapter 3, we briefly mentioned how to define a function with default values. We used : to give a pre-defined value to a given variable,

```
?:
p:v is a pattern object that represents an expression of the
 form p, which, if omitted, should be replaced by v.
```

Here we define a function with two variables, one is supposed to get a symbol with the default value of x and the other an integer with the default value 10. If we don't specify the pattern, we run into problems. The following code shows how we should approach this.

```
g[x_ : x, n_ : 10] := Table[Subscript[x, i], {i, 0, n}]

g[x, 3]
{x_0, x_1, x_2, x_3}

g[y, 7]
{y_0, y_1, y_2, y_3, y_4, y_5, y_6, y_7}

g[]
{x_0, x_1, x_2, x_3, x_4, x_5, x_6, x_7, x_8, x_9, x_10}

g[3]
{3_0, 3_1, 3_2, 3_3, 3_4, 3_5, 3_6, 3_7, 3_8, 3_9, 3_10}

g[x_Symbol : x, n_Integer : 10] := Table[Subscript[x, i], {i, 0, n}]

g[3]
{x_0, x_1, x_2, x_3}

g[u]
{u_0, u_1, u_2, u_3, u_4, u_5, u_6, u_7, u_8, u_9, u_10}
```

12
Recursive functions

A recursive function is a function which calls it-self in its definition (true, pretty confusing!). However, recursive functions arise very naturally. This chapter studies recursive functions and how these functions are handled with ease in *Mathematica*.

Imagine two mirrors are placed (almost) parallel to each other with an apple sitting between them. Then one can see an infinite number of apples in the mirrors. This might give an impression of what a recursive function is. The classic example is Fibonacci numbers. Consider the sequence of numbers starting with the two terms 1 and 1 and continuing with the sum of the two previous terms as the next term in the sequence. Following this rule, one obtains the sequence $1, 1, 2, 3, 5, 8, 13, 21, \cdots$. To define this sequence mathematically, one writes $F_1 = F_2 = 1$ and $F_n = F_{n-1} + F_{n-2}$.

One can use Wolfram *Mathematica*® to define Fibonacci numbers in the exact same way recursively:

```
f[1] = 1; f[2] = 1;
f[n_] := f[n-1] + f[n-2]

f /@ Range[10]
{1, 1, 2, 3, 5, 8, 13, 21, 34, 55}

Fibonacci /@ Range[10]
{1, 1, 2, 3, 5, 8, 13, 21, 34, 55}
```

Now, using f, try to compute the 50th Fibonacci number. This will take a ridiculously long time. What is the problem? The problem will reveal itself if you try to calculate, say, $f[5]$ by hand. By definition, $f[5] = f[4] + f[3]$ thus one needs to calculate $f[4]$ and $f[3]$. Again by definition $f[4] = f[3] + f[2]$ and $f[3] = f[2] + f[1]$. Thus in order to find the value of $f[4]$ one needs to find out $f[3]$ and $f[2]$ and for $f[3]$ one needs to calculate $f[2]$ and $f[1]$. Thus

© Springer International Publishing Switzerland 2015
R. Hazrat, *Mathematica*®: *A Problem-Centered Approach*, Springer Undergraduate
Mathematics Series, DOI 10.1007/978-3-319-27585-7_12

Mathematica is trying to calculate $f[3]$ twice, unnecessarily. This shows that in order to save time, one needs to save the values of the functions in the memory. This has been done in the following code. Compare this with the above.

```
Clear[f]

f[1] = 1; f[2] = 1;
f[n_] := f[n] = f[n - 1] + f[n - 2]

f[50]
12586269025
```

Problem 12.1

Using a recursive definition, write the function

$$e(n) = 1 + \frac{1}{1!} + \frac{1}{2!} + \frac{1}{3!} + \cdots + \frac{1}{n!}.$$

$\Longrightarrow$ Solution.

This is Problem 7.1, in which we used Sum to write the function. In fact we have seen two ways to write this function so far:

```
e[n_] := 1 + Sum[1/k!, {k, 1, n}]

e[10]
9864101/3628800
```

Or use list-based programming as in Chapter 5:

```
Clear[e]
e[n_] := 1 + Plus @@ (1/Range[n]!)

e[10]
9864101/3628800
```

Now we can write the same function but using a recursive method as follows:

```
Clear[e]
e[1] = 1 + 1
2

e[n_] := e[n] = e[n - 1] + 1/n!
e[10]

9864101/3628800
```

Now let us calculate e[5000] using the last function:

```
e[5000]

During evaluation of In[462]:= $RecursionLimit::reclim: Recursion
  depth of 1024 exceeded. >>

Hold[e[3978 - 1] + 1/3978!]
```

This says *Mathematica* has a limitation on the number of recursive evaluations. By default this is 1024 (*Mathematica* would circle around herself up to 1024 times!). If we need to have more iterations, we can change this using `RecursionLimit`. We will change this to 6000 and then calculate $e(5000)$ using all three definitions and we will time it to see which function is the fastest. Notice how `Block` has been used to change the limit of recursion locally for only this computation.

```
$RecursionLimit
1024

Block[{$RecursionLimit = 6000}, Timing[e[5000]][[1]]]

Timing[e[5000]][[1]]
2.609570

Clear[e]
e[n_] := 1 + Sum[1/k!, {k, 1, n}]

Timing[e[5000]][[1]]
2.540101

Clear[e]
e[n_] := 1 + Plus @@ (1/Range[n]!)

Timing[e[5000]][[1]]
0.291022
```

Exercise 12.1

Recently Eric Rowland from Rutgers has come up with the following formula to generate prime numbers.[1] Consider the recursive function described as follows: $a(1) = 7$ and $a(n) = a(n-1)+\gcd(n, a(n-1))$ where gcd is the greatest common divisor (the `GCD` function in *Mathematica*). He then proves that, for any n, $a(n) - a(n-1)$ is either 1 or a prime number. Try out this function and find some prime numbers!

▬ Problem 12.2

Define a function $f : \mathbb{N} \to \mathbb{N}$ as follows: $f(1) = 1, f(2n) = f(n)$ and $f(2n+1) = f(2n) + 1$. Find the smallest n such that $f(n) = 10$.

$\Longrightarrow$ SOLUTION.

---

[1] E. Rowland, A Natural Prime-Generating Recurrence, Journal of Integer Sequences, Vol. 11 (2008), Article 08.2.8

We define f as a function with multiple definitions (see Chapter 11):

```
f[1] = 1;
f[n_?EvenQ] := f[n] = f[n/2]
f[n_?OddQ]  := f[n] = f[n - 1] + 1
```

Now a simple `While` loop can find the smallest n as required by the problem.

```
Module[{n = 1},
 While[f[n] != 10,
   n++];
     n]
```

1023

Exercise 12.2

Define a function $f : \mathbb{N} \to \mathbb{N}$ as follows: $f(1) = 1, f(3) = 3$ and for all $n \in \mathbb{N}$,

$$f(2n) = f(n),$$
$$f(4n + 1) = 2f(2n + 1) - f(n),$$
$$f(4n + 3) = 3f(2n + 1) - 2f(n).$$

Find the number of $n \le 2015$ such that $f(n) = n$.

Problem 12.3

Let the intertwined recursive functions s and t be defined as follows:
$s(1) = 3, s(2) = 5, t(1) = 1$ and

$$s(n) = t(n - 1) - t(n - 2) + 4$$
$$t(n) = t(n - 1) + s(n - 1).$$

Check that $t(100) = 10000$.

$\Longrightarrow$ SOLUTION.

```
s[1] = 3; s[2] = 5; t[1] = 1;

s[n_] := s[n] = t[n - 1] - t[n - 2] + 4

t[n_] := t[n] = t[n - 1] + s[n - 1]
```

Let us first check this for smaller values of n. Notice that $t(4) = t(3) + s(3)$, thus, in order to evaluate $t(4)$, one needs to evaluate $s(3)$ which is by definition $s(3) = t(2) + t(1) + 4$. This shows the definitions of these functions are dependent on each other.

```
t[2]
4

t[3]
9

t[4]
16

t[100]
10000
```

⌐

──── Problem 12.4

Define the function f as follows

$$f(n) = \begin{cases} n - 3 & \text{if } n \geq 1000 \\ f(f(n+6)) & \text{if } n < 1000. \end{cases}$$

Find the value $f(1)$.

⟹ SOLUTION.

The definition of the function consists of two parts, depending on the value of n. Recall, from Section 6.3, that one can use an If statement to define functions of this type:

```
f[n_] := If[n >= 1000, n - 3, f[f[n + 6]]]
```

It is always a very good idea to test the function for some values that one can actually evaluate by hand, just to make sure the code is correct.

```
f[1003]
1000

f[999]
 999

f[1]
997
```

⌐

Here we use recursive programming to solve Problem 8.13.

_____ Problem 12.5

A happy number is a number with the property that if one squares its digits
and adds them together, and then takes the result and squares its digits and
adds them together again and keeps doing this process, one comes down to the
number 1. Find all the happy ages, i.e., happy numbers up to 100.

⟹ SOLUTION.

```
f[1] = 1
f[4] = 4

f[n_] := f[Plus @@ ( IntegerDigits[n]^2)]

Select[Range[100], f[#] == 1 &]
{1, 7, 10, 13, 19, 23, 28, 31, 32, 44, 49, 68, 70, 79, 82,
86, 91, 94, 97, 100}
```

This does not seem to be a good definition for a happy number, as according
to this definition and the above result there are only four years between 30 and
retirement that a person is happy!

One can re-write many codes which have a repetitive nature in the form of
recursion. Recall the Collatz function from Problem 11.2. One can write the
function as follows

```
f[1]=1
f[n_Integer?EvenQ] := f[n/2]
f[n_Integer] := f[3n + 1]
```

Then if one applies f to any number one should get 1. (If not, then one has
solved an 80-year-old conjecture in the negative!)

Exercise 12.3

Using a recursive definition, write the function

$$f(k) = x + \frac{x^3}{1 \times 2 \times 3} + \frac{x^5}{1 \times 2 \times 3 \times 4 \times 5} + \cdots + \frac{x^k}{1 \times 2 \times \cdots \times k}$$

(k is odd).

Recall continued fractions from Problem 8.4. We will calculate them here
using recursive functions.

―――― Problem 12.6

Given non-negative integers $c_1, c_2, \ldots, c_m$, where $c_i \geq 0$ for $i \geq 2$, define

$$[c_1, c_2, \ldots, c_m] = c_1 + \cfrac{1}{c_2 + \cfrac{1}{\cdots + \cfrac{1}{c_m}}}.$$

Write a recursive function to accept a list $\{c_1, c_2, \ldots, c_m\}$ and calculate the continued fraction $[c_1, c_2, \ldots, c_m]$.

Furthermore, for the list $\{c_1, c_2, \ldots, c_m\}$ define the integers p_m and q_m recursively as follows:

$$p_1 = c_1, \quad p_2 = c_2 c_1 + 1, \qquad p_m = c_m p_{m-1} + p_{m-2}$$
$$q_1 = 1, \quad q_2 = c_2, \qquad q_m = c_m q_{m-1} + q_{m-2}.$$

Then show that

$$[c_1, c_2, \ldots, c_m] = p_m/q_m.$$

$\Longrightarrow$ SOLUTION.

Once the code is written, it is easy to understand the logic of the code. Here is the code!

```
f[s_List] := s[[1]] /; Length[s] == 1
f[s_List] := s[[1]] + 1/f[Rest[s]]


f[{1, 2, 3}]
10/7
```

Now we write p_m and q_m as recursive functions. Since these two functions are defined is exactly the same way and the only difference is their initial values, we will define them in one go. We define the function pq[s_List,1] for p and pq[s_List,2] for q and assign the initial values to them separately.

```
pq[s_List?(Length[#] == 1 &), 1] := s[[1]];
pq[s_List?(Length[#] == 2 &), 1] := s[[1]] s[[2]] + 1

pq[s_List?(Length[#] == 1 &), 2] := 1 ;
pq[s_List?(Length[#] == 2 &), 2] := s[[2]]
```

Now we define the main body of the function for p and q as follows:

```
pq[s_List?(Length[#] >= 3 &), n_] :=
s[[-1]] pq[Drop[s, -1], n] + pq[Drop[s, -2], n]
```

Here is a test showing that both approaches give us the same result.

```
s = Range[24, 46];

f[s]== pq[s, 1]/pq[s, 2]
True
```

└─────────────────────

────── Problem 12.7

The polynomials $P_n(x, y)$ for $n = 1, 2, \cdots$ are defined by $P_1(x, y) = 1$ and

$$P_{n+1}(x, y) = (x + y - 1)(y + 1)P_n(x, y + 2) + (y - y^2)P_n(x, y).$$

Check first that $P_2(x, y) = xy + x + y - 1$. Then investigate that, for any n, $P_n(x, y) = P_n(y, x)$.

$\Longrightarrow$ SOLUTION.

First we define the recursive function:

```
p[1, x_, y_] = 1;

p[n_, x_, y_] := p[n, x, y] =
(x + y - 1) (y + 1) p[n - 1, x, y + 2] + (y - y^2) p[n -1, x, y]
```

We check whether the definition is correct.

```
p[2, x, y]
y - y^2 + (1 + y) (-1 + x + y)

Simplify[%]
-1 + x + y + x y
```

The definition is correct. Of course we cannot expect *Mathematica* to be able to check $P_n(x, y) = P_n(y, x)$ for undefined n, so we check this for some instances of n.

```
Simplify[p[3, x, y] == p[3, y, x]]
True

Simplify[p[6, x, y] == p[6, y, x]]
True

Simplify[p[16, x, y] == p[16, y, x]]
True
```

└─────────────────────

———— Problem 12.8

For integers a and b, if $a = bq + r$, where $q, r \in \mathbb{Z}$, then $\gcd(a, b) = \gcd(b, r)$. Using this recursively, write a function to calculate the greatest common divisor of two non-negative integers.

$\Longrightarrow$ SOLUTION.

This is called Euclid's algorithm for finding the greatest common divisor. It is based on the fact that for integers a and b there are unique integers q and r such that $a = bq + r$, where $0 \le r \le |b|$ (see Tips on page 37).

We first work out an example by hand to show the recursive nature of this approach. To find the greatest common divisor of 3780 and 132, we can write:

$$
\begin{aligned}
3780 &= 132 \times 28 + 84 & \gcd(3780, 132) &= \gcd(132, 84) \\
132 &= 84 \times 1 + 48 & \gcd(132, 84) &= \gcd(84, 48) \\
84 &= 48 \times 1 + 36 & \gcd(84, 48) &= \gcd(48, 36) \\
48 &= 36 \times 1 + 12 & \gcd(48, 36) &= \gcd(36, 12) \\
36 &= 12 \times 3 + 0 & \gcd(36, 12) &= \gcd(12, 0) = 12.
\end{aligned}
$$

Thus $\gcd(3780, 132) = 12$.

We need to write a function with multiple definitions (see Chapter 11).

```
gcd[a_, 0] := a;
gcd[0, b_] := b;
gcd[a_, b_] := gcd[b, Mod[a, b]] /; a >= b;
gcd[a_, b_] := gcd[a, Mod[b, a]] /; a < b

gcd[3780, 132]
12
```

Of course the built-in function GCD can easily calculate the greatest common divisor.

```
GCD[3780, 132]
12

FactorInteger[3780]
{{2, 2}, {3, 3}, {5, 1}, {7, 1}}

FactorInteger[132]
{{2, 2}, {3, 1}, {11, 1}}
```

Exercise 12.4

Use `PolynomialQuotient` and `PolynomialRemainder` to write Euclid's algorithm for finding the greatest common divisors of polynomials similar to Problem 12.8.

Exercise 12.5

Define the following function in *Mathematica*,

$$f(m, -m) = x_m$$
$$f(m, n) = f(m - 1, n) + f(m, n - 1),$$

where $m, n \in \mathbb{Z}$, and show that

$$f(2, 5) = x_{-5} + 7x_{-4} + 21x_{-3} + 35x_{-2} + 35x_{-1} + 21x_0 + 7x_1 + x_2.$$

Further define an *involution* map ϕ sending x_k to x_{-k} and show that $\phi(f(2, 5)) = f(5, 2)$.

13

Linear algebra

Computations with matrices are tedious jobs. One can easily
define vectors and matrices in *Mathematica* and perform all
the standard procedures of linear algebra using ready-to-use
Mathematica functions. This chapter looks at these facilities.

13.1 Vectors

One of the questions asked in Chapter 4 was as follows: Given $\{x_1, x_2, \cdots, x_n\}$
and $\{y_1, y_2, \cdots, y_n\}$, how can one produce $\{x_1 + y_1, x_2 + y_2, \cdots, x_n + y_n\}$?
The answer is if one considers lists as *vectors*, i.e., $\mathbf{x} = \{x_1, x_2, \cdots, x_n\}$ and
$\mathbf{y} = \{y_1, y_2, \cdots, y_n\}$, where $\mathbf{x}$ and $\mathbf{y}$ are two vectors of dimension n, then the
sum of vectors, $\mathbf{x} + \mathbf{y}$ is what we want. Then, as you might also guess, $\mathbf{xy}$ would
produce $\{x_1 y_1, x_2 y_2, \cdots, x_n y_n\}$, i.e., all arithmetical operations performed on
lists are component-wise. However, there is another product in the setting of
vectors, namely the *inner product*, which is defined as

$$\mathbf{x}.\mathbf{y} = x_1 y_1 + x_2 y_2 + \cdots + x_n y_n.$$

The following shows how Wolfram *Mathematica*® handles these different operations.

```
xv = x# & /@ Range[5]
```

```
{x₁, x₂, x₃, x₄, x₅}
```

© Springer International Publishing Switzerland 2015
R. Hazrat, *Mathematica®: A Problem-Centered Approach*, Springer Undergraduate
Mathematics Series, DOI 10.1007/978-3-319-27585-7_13

```
yv = y# & /@ Range[5]
```

$\{y_1, y_2, y_3, y_4, y_5\}$

```
xv + yv
```

$\{x_1 + y_1, x_2 + y_2, x_3 + y_3, x_4 + y_4, x_5 + y_5\}$

```
xv yv
```

$\{x_1\,y_1, x_2\,y_2, x_3\,y_3, x_4\,y_4, x_5\,y_5\}$

```
xv * yv
```

$\{x_1\,y_1, x_2\,y_2, x_3\,y_3, x_4\,y_4, x_5\,y_5\}$

```
xv.yv
```

$x_1\,y_1 + x_2\,y_2 + x_3\,y_3 + x_4\,y_4 + x_5\,y_5$

```
Norm[xv]
```

$$\sqrt{\mathrm{Abs}[x_1]^2 + \mathrm{Abs}[x_2]^2 + \mathrm{Abs}[x_3]^2 + \mathrm{Abs}[x_4]^2 + \mathrm{Abs}[x_5]^2}$$

13.2 Matrices

It is known that matrix calculation is a tedious job. It will take well over 10 minutes to multiply

$$\begin{pmatrix} 2 & -3 & 13 & -4 & 8 \\ 12 & 1 & -18 & -4 & 2 \\ 18 & 21 & 10 & 0 & 9 \\ 8 & -12 & -4 & 0 & -3 \\ 15 & -7 & 2 & 4 & 2 \end{pmatrix} \times \begin{pmatrix} 11 & 34 & -21 & 0 & -43 \\ 12 & -33 & 9 & -12 & 7 \\ 16 & -7 & -43 & 84 & 3 \\ 4 & 9 & 12 & -1 & -54 \\ 7 & 22 & -5 & 23 & 0 \end{pmatrix}$$

by hand only to obtain a wrong answer!

One can easily enter a matrix into *Mathematica* by choosing Insert: Table/Matrix from the menu. If we assign A to the first matrix and B to the second then

A.B // MatrixForm

$$\begin{pmatrix} 234 & 216 & -716 & 1316 & 148 \\ -146 & 509 & 473 & -1474 & -347 \\ 673 & 47 & -664 & 795 & -597 \\ -141 & 630 & -89 & -261 & -440 \\ 143 & 807 & -426 & 294 & -904 \end{pmatrix}$$

Det[A]

354 956

Note that, similar to multiplication of vectors, one should use Dot multiplication for multiplying matrices, i.e., **A.B**. One uses the function **MatrixForm** to obtain the result in, well, matrix form! Otherwise one gets a list of vectors. The command **Det** calculates the determinant of the matrix.

Even more impressive is how easily *Mathematica* computes the inverse of this matrix. Try

Inverse[A] // MatrixForm

$$\begin{pmatrix} -\dfrac{828}{88\,739} & -\dfrac{375}{88\,739} & \dfrac{3310}{88\,739} & \dfrac{6670}{88\,739} & \dfrac{1203}{88\,739} \\ -\dfrac{2462}{88\,739} & \dfrac{472}{88\,739} & \dfrac{2983}{88\,739} & \dfrac{113}{88\,739} & -\dfrac{2934}{88\,739} \\ \dfrac{1077}{88\,739} & -\dfrac{4335}{88\,739} & \dfrac{2768}{88\,739} & \dfrac{6114}{88\,739} & \dfrac{3258}{88\,739} \\ -\dfrac{692}{12\,677} & -\dfrac{335}{50\,708} & \dfrac{2959}{50\,708} & \dfrac{10\,099}{50\,708} & \dfrac{4787}{25\,354} \\ \dfrac{6204}{88\,739} & \dfrac{6668}{88\,739} & \dfrac{6796}{88\,739} & -\dfrac{20\,397}{88\,739} & \dfrac{12\,872}{88\,739} \end{pmatrix}$$

One can generate a matrix by using **Array**, as the following example demonstrates:

m = Array[#1 - #2 &, {3, 4}]

{{0, -1, -2, -3}, {1, 0, -1, -2}, {2, 1, 0, -1}}

m // MatrixForm

$$\begin{pmatrix} 0 & -1 & -2 & -3 \\ 1 & 0 & -1 & -2 \\ 2 & 1 & 0 & -1 \end{pmatrix}$$

Here, **Array** acts as a nested loop (see Section 8.2). That is, the first variable (here **#1**) runs from 1 to 3 while the second variable (**#2**) runs from 1 to 4. This is equivalent to

```
Table[i - j, {i, 1, 3}, {j, 1, 4}]

{{0, -1, -2, -3}, {1, 0, -1, -2}, {2, 1, 0, -1}}
```

However, it is much more convenient and more readable to use **Array** when generating matrices.

━━━ Problem 13.1

Define a 3×2 matrix (a_{ij}) with entries $a_{ij} = i - j$.

$\Longrightarrow$ SOLUTION.
    We will use **Array** to generate the matrix.

```
Array[#1 - #2 &, {3, 2}] // MatrixForm
```

$$\begin{pmatrix} 0 & -1 \\ 1 & 0 \\ 2 & 1 \end{pmatrix}$$

Note that in the **Array**, **#1** stands for i and **#2** for j.

A common mistake here is to use the command **MatrixForm** in the definition of a matrix. Suppose we want to define an $n \times n$ matrix (a_{ij}) with entries $a_{ij} = i - j^i$. It is very tempting to define the function as

```
m[n_] := Array[#1 - #2 ^ #1 &, {n, n}] // MatrixForm
```

Everything looks fine and we get the matrix form of what we want.

```
m[3]
```

$$\begin{pmatrix} 0 & -1 & -2 \\ 1 & -2 & -7 \\ 2 & -5 & -24 \end{pmatrix}$$

However, let us calculate the determinant of this function.

```
Det[m[3]]
```

$$\text{Det}\left[\begin{pmatrix} 0 & -1 & -2 \\ 1 & -2 & -7 \\ 2 & -5 & -24 \end{pmatrix}\right]$$

The reason is, using `MatrixForm`, our output is no longer a two-dimensional list that *Mathematica* interprets as a matrix but a table of data.

—— Problem 13.2

Write a function to check that, for any n, the following identity holds:

$$\det\begin{pmatrix} 1 & 1 & 1 & \cdots & 1 & 1 \\ b_1 & a_1 & a_1 & \cdots & a_1 & a_1 \\ b_1 & b_2 & a_2 & \cdots & a_2 & a_2 \\ \vdots & \vdots & \vdots & & \vdots & \vdots \\ b_1 & b_2 & b_3 & \cdots & b_n & a_n \end{pmatrix} = (a_1 - b_1)(a_2 - b_2)\cdots(a_n - b_n)$$

⟹ SOLUTION.

Here is the definition of the matrix using `Array`.

```
m[n_] := Array[
  Which[
    #1 == 1, 1,
    #1 ≤ #2, a#1-1,
    True, b#2] &,
  {n + 1, n + 1}]
```

Let's check that m actually produces matrices of the above form.

```
m[5] // MatrixForm
```

$$\begin{pmatrix} 1 & 1 & 1 & 1 & 1 & 1 \\ b_1 & a_1 & a_1 & a_1 & a_1 & a_1 \\ b_1 & b_2 & a_2 & a_2 & a_2 & a_2 \\ b_1 & b_2 & b_3 & a_3 & a_3 & a_3 \\ b_1 & b_2 & b_3 & b_4 & a_4 & a_4 \\ b_1 & b_2 & b_3 & b_4 & b_5 & a_5 \end{pmatrix}$$

```
m2[n_] := Product[a_i - b_i, {i, 1, n}]

m2[7]

(a₁ - b₁) (a₂ - b₂) (a₃ - b₃) (a₄ - b₄) (a₅ - b₅) (a₆ - b₆) (a₇ - b₇)

Simplify[Det[m[7]] == m2[7]]

True
```

Problem 13.3

Write a function to check that, for any n, the following identity holds:

$$\det \begin{pmatrix} x & a_1 & a_2 & \cdots & a_n \\ a_1 & x & a_2 & \cdots & a_n \\ a_1 & a_2 & x & \cdots & a_n \\ \vdots & \vdots & \vdots & & \vdots \\ a_1 & a_2 & a_3 & \cdots & x \end{pmatrix} = (x + a_1 + \cdots + a_n)(x - a_1) \cdots (x - a_n)$$

$\Longrightarrow$ SOLUTION.
    The solution is left to the reader this time!

Exercise 13.1

Write a function to accept a matrix A_{nn} and produce the $n^2 \times n^2$ matrix B as follows,

$$\begin{pmatrix} \begin{pmatrix} a_{11} & 0 & 0 \\ 0 & \ddots & 0 \\ 0 & 0 & a_{11} \end{pmatrix} & \begin{pmatrix} a_{12} & 0 & 0 \\ 0 & \ddots & 0 \\ 0 & 0 & a_{12} \end{pmatrix} & \cdots & \begin{pmatrix} a_{1n} & 0 & 0 \\ 0 & \ddots & 0 \\ 0 & 0 & a_{1n} \end{pmatrix} \\ \vdots & \vdots & \vdots & \vdots \\ \begin{pmatrix} a_{n1} & 0 & 0 \\ 0 & \ddots & 0 \\ 0 & 0 & a_{n1} \end{pmatrix} & \cdots & \cdots & \begin{pmatrix} a_{nn} & 0 & 0 \\ 0 & \ddots & 0 \\ 0 & 0 & a_{nn} \end{pmatrix} \end{pmatrix}$$

Then show that $\det(A)^n = \det(B)$.

Exercise 13.2

Define a matrix as follows:

$$
d(n) = \begin{pmatrix}
1 & 2 & \cdots & n \\
n+1 & n+2 & \cdots & 2n \\
\vdots & \vdots & \vdots & \vdots \\
\cdots & \cdots & \cdots & n^2
\end{pmatrix}.
$$

Check that, for any $n > 2$, $\det(d(n)) = 0$.

The following is a nice problem demonstrating the use of pattern matching in *Mathematica* for solving problems.

——— Problem 13.4

Let A and B be 3×3 matrices. Show that $(ABA^{-1})^5 = AB^5 A^{-1}$.

$\Longrightarrow$ SOLUTION.

Let us first take the naive approach. We define two arbitrary matrices and, using *Mathematica*, we will multiply them and check whether both sides give the same result.

```
(A = Array[x#1,#2 &, {3, 3}]) // MatrixForm
```

$$
\begin{pmatrix}
x_{1,1} & x_{1,2} & x_{1,3} \\
x_{2,1} & x_{2,2} & x_{2,3} \\
x_{3,1} & x_{3,2} & x_{3,3}
\end{pmatrix}
$$

```
(B = Array[y#1,#2 &, {3, 3}]) // MatrixForm
```

$$
\begin{pmatrix}
y_{1,1} & y_{1,2} & y_{1,3} \\
y_{2,1} & y_{2,2} & y_{2,3} \\
y_{3,1} & y_{3,2} & y_{3,3}
\end{pmatrix}
$$

```
IA = Inverse[A];
IB = Inverse[B];
```

Now we check the equality for $n = 3$ and time it:

```
Timing[d = A.B.IA.A.B.IA.A.B.IA; d1 = A.B.B.B.IA;
 Simplify[d == d1]]
```

{61.9128, True}

This already takes a long time. One can easily prove by induction that $(ABA^{-1})^n = AB^n A^{-1}$ for any positive integer n. For example, for $n = 2$ we have $(ABA^{-1})^2 = ABA^{-1}ABA^{-1} = ABBA^{-1} = AB^2 A^{-1}$. This reveals a pattern. Namely, we can easily cancel A with A^{-1} if they are adjacent to each other. We introduce this to *Mathematica* and try to check the equality this way. This is very similar in nature to Problem 11.5.

```
ml[x___, y_, z_, t___] := ml[x, t] /; Simplify[y.z == IdentityMatrix[3]]

ml[r__] := Apply[Dot, {r}]

ml[] = 1;
```

```
Timing[Simplify[ml[A, B, IA, A, B, IA, A, B, IA] == A.B.B.B.IA]]
```

{0.005241, True}

♣ TIPS

– For a square matrix, Eigenvalues and Eigenvectors determine these invariants of a matrix.

Exercise 13.3

Let $b_1, b_2, b_3, b_4, \ldots$ denote the sequence defined by $b_1 = 1, b_2 = 1, b_3 = 2$ and

$$\det \begin{pmatrix} b_1 & b_1 + b_2 & 1 \\ b_3 & b_3 + b_4 & 0 \\ 0 & 0 & 1 \end{pmatrix} = \det \begin{pmatrix} b_2 & b_2 + b_3 & 1 \\ b_4 & b_4 + b_5 & 0 \\ 0 & 0 & 1 \end{pmatrix}$$
$$= \det \begin{pmatrix} b_3 & b_3 + b_4 & 1 \\ b_5 & b_5 + b_6 & 0 \\ 0 & 0 & 1 \end{pmatrix} = \cdots = 1.$$

Check that, for any given i, b_i is an integer.

14
Graphics

This chapter looks at graphics: how to use *Mathematica* to plot complex graphs. We first look at two-dimensional graphs and then concentrate on three-dimensional graphs. *Mathematica* offers similar tools in both two- and three-dimensional cases.

Graphics is one of the strongest features of Wolfram *Mathematica*®. One can use *Mathematica* to create the plot of a very complex function. We first look at two-dimensional graphs and then concentrate on three-dimensional graphs. *Mathematica* offers similar tools in both two- and three-dimensional cases.

14.1 Two-dimensional graphs

It is always very helpful to present the behaviour of a function or an equation by plotting its graph. For functions with one variable, this can be done by two-dimensional graphs. Functions can come in different forms, and *Mathematica* has specific commands to handle each case. The following table shows which command is suitable for different formats of functions.

Function	Example	Graphics command
$y = f(x)$	$y = \sin(x)/x$	`Plot`
$x = f(t), y = g(t)$	$x = \sin(3t), y = \cos(4t)$	`ParametricPlot`
$f(x,y) = 0$	$x^4 - (x^2 - y^2) = 0$	`ContourPlot`
$f(x,y) \geq 0$	$x^4 + (x - 2y^2) > 0$	`RegionPlot`
$r = f(\theta)$	$r = 3\cos(6\theta)$	`PolarPlot`

© Springer International Publishing Switzerland 2015
R. Hazrat, *Mathematica*®: *A Problem-Centered Approach*, Springer Undergraduate
Mathematics Series, DOI 10.1007/978-3-319-27585-7_14

▬▬▬ Problem 14.1

Plot the graphs of the following functions:

$$f(x) = \sin(x)/x$$
$$x = \sin(3t), y = \cos(4t)$$
$$x^4 - (x^2 - y^2) = 0$$
$$x^4 + (x - 2y^2) > 0$$
$$r = 3\cos(6\theta).$$

$\Longrightarrow$ SOLUTION.

As the table on page 205 shows, to plot the graph of $\sin(x)/x$ we need to use Plot.

```
Plot[Sin[x]/x, {x, 0, 10 Pi}]
```

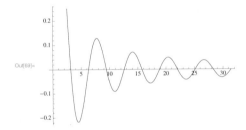

Figure 14.1 $f(x) = \sin(x)/x$

For $x = \sin(3t), y = \cos(4t)$, the command ParametricPlot is available.

```
ParametricPlot[{Sin[3 t], Cos[4 t]}, {t, 0, 2 Pi}]
```

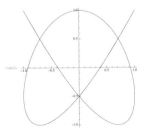

Figure 14.2 $x = \sin(3t), y = \cos(4t)$

To find the graph of all the points (x, y) which satisfy the equation $x^4 - (x^2 - y^2) = 0$, one uses `ContourPlot`.

```
ContourPlot[x^4 - (x^2 - y^2) == 0, {x, -1, 1}, {y, -1, 1}]
```

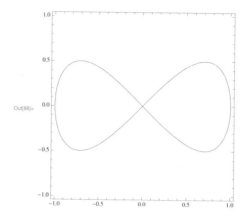

Figure 14.3 $x^4 - (x^2 - y^2) = 0$

To obtain the region, namely all the points (x, y) which satisfy the inequality $x^4 + (x - 2y^2) > 0$, we have `RegionPlot`.

```
RegionPlot[x^4 + (x - 2 y^2) >= 0, {x, -2, 2}, {y, -2, 2}]
```

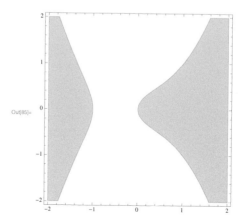

Figure 14.4 $x^4 + (x - 2y^2) > 0$

Finally to plot the polar graph of $r = 3\cos(6\theta)$, we need to use `PolarPlot`.

```
PolarPlot[3 Cos[6 t], {t, 0, 2 Pi}]
```

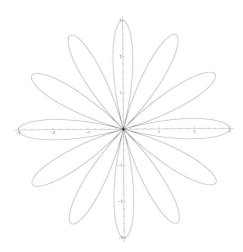

Figure 14.5 $r = 3\cos(6\theta)$

Note that in all the cases, we need to specify an interval over which we would like to plot the graph.

∟_____

━━━ Problem 14.2

Show that there are 11 solutions to the equation $2\cos^2(2x) - \sin(x/2) = 1$ in the interval $[0, 3\pi]$.

⟹ SOLUTION.
    The best approach here is to plot the graph of $2\cos^2(2x) - \sin(x/2) - 1$ and count the number of times the graph crosses the x-axis.

```
Plot[2 Cos[2 x]^2 - Sin[x/2] - 1, {x, 0, 3 Pi}]
```

From the graph (Fig. 14.6), it seems there are 11 solutions to this equation. However, there might be some concern around the region $3.1 \leq x \leq 3.2$, as it is not clear whether the graph actually touches the x-axis, that is, whether there is a root there or not. So we focus on that region.

```
Plot[2 Cos[2 x]^2 - Sin[x/2] - 1, {x, 3.1, 3.2}]
```

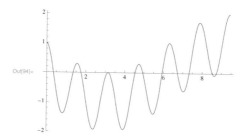

Figure 14.6 $2\cos^2(2x) - \sin(x/2) - 1$

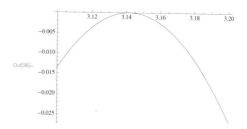

Figure 14.7 $2\cos^2(2x) - \sin(x/2) - 1$ in the region $3.1 \le x \le 3.2$

This graph (Fig. 14.7) makes it clear that in fact there is a root in this region. We can use *Mathematica*'s ability to find roots to make sure this is the case. (For more on this, see Chapter 15.)

```
FindRoot[2 Cos[2 x]^2 - Sin[x/2] - 1, {x, 3.14}]
{x -> 3.14159}
```

Problem 14.3

Observe the behavior of the graph $\sin(x) - \cos(nx)$ between 0 and π as n changes from 1 to 100.

$\Longrightarrow$ SOLUTION.

In order to record the changes in the behaviour of the graph as n runs from 1 to 100 we use Manipulate. The code is easy, just wrap the command Manipulate around Plot and let n run from 1 to 100.

```
Manipulate[Plot[Sin[x] - Cos[n x], {x, 0, Pi}], {n, 1, 100, 1}]
```

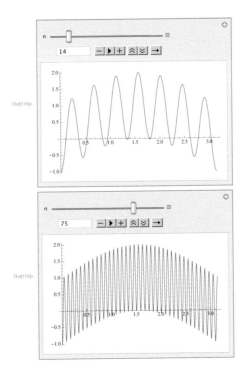

Figure 14.8 The graph of $\sin(x) - \cos(nx)$ for $n = 14$ and $n = 75$

Problem 14.4

Draw the butterfly curve, discovered by Temple H. Fay, given by

$$x(t) = \sin(t)\left(e^{\cos(t)} - 2\cos(4t) - \sin^5(t/12)\right)$$
$$y(t) = \cos(t)\left(e^{\cos(t)} - 2\cos(4t) - \sin^5(t/12)\right)$$

$\Longrightarrow$ SOLUTION.

This is similar to the second equation in Problem 14.1. It is a parametric equation and thus calls for `ParametricPlot`

```
x[t_] := Sin[t] (E^Cos[t] - 2 Cos[4 t] - Sin[t/12]^5)

y[t_] := Cos[t] (E^Cos[t] - 2 Cos[4 t] - Sin[t/12]^5)

ParametricPlot[{x[t], y[t]}, {t, -50, 50}]
```

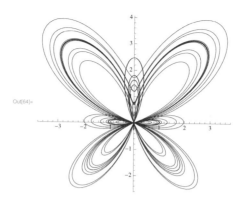

Figure 14.9   The butterfly curve

Mathematica can plot the graphs of several equations simultaneously. For this one introduces the equations into `Plot` by using a list containing all the equations.

Problem 14.5

Plot the graphs of the functions $\sin(\frac{1}{x^2-x})$ and $\frac{1.5}{x}$ in the range $[0, \pi]$.

$\Longrightarrow$ SOLUTION.

One can plot the graph of each function separately and then use the command `Show` to show two graphs combined. However, one can also plot two graphs simultaneously (see Fig. 14.10).

```
Plot[{Sin[1/(x^2 - x)], 1.5/x}, {x, 0, Pi}]
```

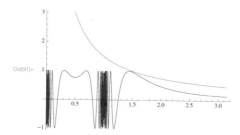

Figure 14.10 $\sin(\frac{1}{x^2-x})$ and $\frac{1.5}{x}$

One issue here is that it is not clear which graph belongs to which equation. One does have access to all aspects of graphs and one can control all parameters which alter the graph of a function. If you type ??Plot, you will see you can change many parameters in the graph. Here are just a few samples, AxesStyle->, Background->, FillingStyle->, FormatType:>, Frame->, FrameLabel->, FrameStyle->, PlotStyle->. For examples on how one can change these options, see *Mathematica* Help on Plot.

Using this, one can produce professional-looking graphs. Here is one attempt to make the above graph look more professional!

```
Plot[{Sin[1/(x^2 - x)], 1.5/x}, {x, 0, Pi}, Frame -> True,
  FrameStyle -> Thick, FrameLabel -> {x - axis, y - axis},
  PlotStyle -> {{Thick}, {Thick, Dashed}}, Background -> Gray,
  PlotLabel -> For Demonstration]
```

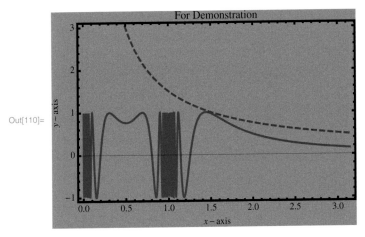

Figure 14.11 $\sin(\frac{1}{x^2-x})$ and $\frac{1.5}{x}$

One of a very useful adds on to Plot is the command Epilog. The following examples shows how Epilog can be used to add more items to a plot.

```
?Epilog
Epilog is an option for graphics functions that gives a list
of graphics primitives to be rendered after the main part
of the graphics is rendered
```

```
Plot[Sin[x^2], {x, -Pi, Pi},
Epilog -> {PointSize[Medium],
  Point[Table[{x, Sin[x^2]}, {x, -3, 3, 0.3}]]}]
```

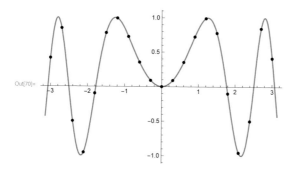

Combining the above code with `Manipulate` to increase the points generated by $\sin(x^2)$ one can create a very interesting demonstration.

```
Manipulate[Plot[Sin[x^2], {x, -Pi, Pi},
Epilog -> {PointSize[Medium], Red,
  Point[Table[{x, Sin[x^2]}, {x, -3, 3, i}]]}], {i, 2, 0.05}]
```

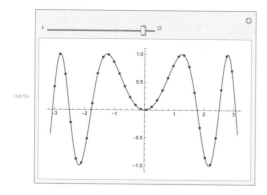

Exercise 14.1

Consider the function

$$f(x) = \frac{|x|}{\sqrt{c - x^2}}.$$

The curves arising from this equation for various c are called bullet-nose curves. Using `Manipulate`, increase c and obtain the graph of the function.

Exercise 14.2

Plot the graph

$$f(x) = \sin(x) - e^{-\sum_{k=1}^{30} \frac{\cos(kx)}{k}} + e^{-\sum_{k=1}^{50} \frac{\sin(kx)}{k}}$$

for x ranging over the interval $[0, 3\pi]$. Find the maximum value that $f(x)$ takes in the interval $[0, 3\pi]$. (Hint, see `Maximize` and `NMaximize`.)

Exercise 14.3

Plot the graph of the expression

$$x(2\pi - x) \sum_{n=1}^{50} \frac{\sin(nx)}{x}$$

for $0 \leq x \leq \pi$.

Exercise 14.4

Plot the graph of

$$x(t) = 4\cos(-11t/4) + 7\cos(t)$$
$$y(t) = 4\sin(-11t/4) + 7\sin(t)$$

for $0 \leq t \leq 14\pi$.

Exercise 14.5

Plot the graph of

$$x(t) = \cos(t) + 1/2\cos(7t) + 1/3\sin(17t)$$
$$y(t) = \sin(t) + 1/2\sin(7t) + 1/3\cos(17t)$$

for $0 \leq t \leq 14\pi$.

━━━━ Problem 14.6

Plot the graph of the function

$$f(x) = \begin{cases} -x, & \text{if } |x| < 1 \\ \sin(x), & \text{if } 1 \leq |x| < 2 \\ \cos(x), & \text{otherwise.} \end{cases} \qquad (14.1)$$

⟹ SOLUTION.

We defined this function in Problem 6.7, using `Which` and `Piecewise`. Plotting the graph is a no-brainer, we need to plug the function $f(x)$ into the command `Plot`. Let us do so, for both definitions of the function $f(x)$.

```
f[x_] := Which[
  Abs[x] < 1, -x,
  1 <= Abs[x] < 2, Sin[x],
  True, Cos[x] ]

Plot[f[x], {x, -4, 4}]
```

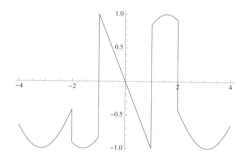

Figure 14.12 Graph of $f(x)$ using `Which`

There is also another way to define this function using the command `Piecewise`.

```
g[x_] := Piecewise[{{-x, Abs[x] < 1}, {Sin[x], 1 <= Abs[x] < 2}},
  Cos[x]]

Plot[g[x], {x, -4, 4}]
```

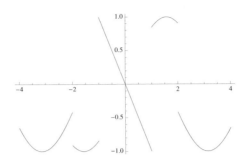

Figure 14.13 Graph of $f(x)$ using `Piecewise`

Comparing the two graphs clearly shows that the correct definition of the function is to use `Piecewise`, where *Mathematica* does not interpret the function as a continuous function.

♣ TIPS

– *Mathematica* produces the graph of a function based on the number of sample points it generates. For more involved functions and higher quality graphs increase `PlotPoints` and `MaxRecursive` (see Problem 14.12).

– To arrange several plots next to each other, use `GraphicsGrid`, `GraphicsRow` and `GraphicsColumn` (see Problem 14.9 for an example using `GraphicsGrid`).

Exercise 14.6

Plot the graphs of the functions $2\exp^{-x^2}$ and $\cos(\sin(x) + \cos(x))$ between $[-\pi, \pi]$.

Exercise 14.7

Plot the graph of $|\ 3x^2 + xy^2 - 12\ | = |\ x^2 - y^2 + 4\ |$.

Exercise 14.8

Plot the graph of the inequality $|\ x^2 + y\ | \leq |\ y^2 + x\ |$.

Exercise 14.9

Plot the graph of $4(x^2 + y^2 - x)^3 - 27(x^2 + y^2)^2 = 0$.

Problem 14.7

Plot the graphs of $x^4 - (x^2 - y^2) = 0$ and $y^4 - (y^2 - x^2) = 0$. Using `Manipulate`, observe how the coefficients $0 \leq a \leq 1$ and $0 \leq b \leq 1$ rescale the graph in

$$(ax)^4 - ((ax)^2 - (ay)^2) = 0,$$
$$(by)^4 - ((by)^2 - (bx)^2) = 0.$$

$\Longrightarrow$ SOLUTION.

From the table on page 205 it is clear that one needs to use `ContourPlot` here. As in the case of `Plot`, one can feed a list of equations into `ContourPlot` to have a graph of several equations simultaneously (see Problem 14.5). Here is the code with `Manipulate`.

```
Manipulate[
 ContourPlot[{(a x)^4 - ((a x)^2 - (a y)^2) == 0,
  (b y)^4 - ((b y)^2 - (b x)^2) == 0}, {x, -1, 1}, {y, -1, 1}],
  {a, 1, 2}, {b, 1, 2}]
```

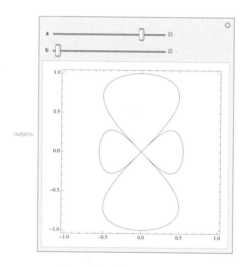

Figure 14.14 $x^4 - (x^2 - y^2) = 0$ and $y^4 - (y^2 - x^2) = 0$

Problem 14.8

Consider the first 10000 digits of $\sqrt{2}$ and present them as a "random walk" by converting them in base 4, representing 4 directions (up, down, left and right). We know that $\sqrt{2}$ is an irrational number and irrational numbers have decimal expansions that neither terminate nor become periodic. Write a code to produce this random walk. Try this code with $\sqrt{3}$, $\sqrt{6}$ and $\sqrt{13}$. Is there any comparison one can make among these numbers?

$\Longrightarrow$ SOLUTION.

   I found this problem and its slick approach in a blog on the Internet.[1] The code uses FoldList to move the object in different directions consecutively, thus creating a random walk. This idea is used in many similar situations (see, e.g, §2.3 in [7] and §11.2 in [3]). Let us start step by step. We first get the first 20 digits of $\sqrt{2}$ and convert them to base 4. All this can be done by RealDigits.

```
x = N[Sqrt[2], 20]
1.4142135623730950488

walk = First@RealDigits[x, 4]
{1, 1, 2, 2, 2, 0, 0, 2, 1, 3, 2, 1, 2, 1, 2, 1, 3, 3, 3, 0,
 3, 2, 3, 3, 0, 3, 0, 2, 1, 0, 0, 2, 0}
```

Then when we encounter a zero, we move the object one step up, that is, we add $(0,1)$ to the point; when we encounter a 1, we move the object one step to the right, that is we add $(1,0)$ to the point and so on:

```
{{0, 1}, {1, 0}, {0, -1}, {-1, 0}}[[# + 1]] & /@ walk

{{1, 0}, {1, 0}, {0, -1}, {0, -1}, {0, -1}, {0, 1}, {0,
  1}, {0, -1}, {1, 0}, {-1, 0}, {0, -1}, {1, 0}, {0, -1},
 {1, 0}, {0, -1}, {1, 0}, {-1, 0}, {-1, 0}, {-1, 0}, {0, 1},
 {-1, 0}, {0, -1}, {-1, 0}, {-1, 0}, {0, 1}, {-1, 0}, {0, 1},
 {0, -1}, {1, 0}, {0, 1}, {0, 1}, {0, -1}, {0, 1}}
```

   The next step is to take these movements into account consecutively. For this FoldList is an excellent tool to use (see Section 8.4).

```
FoldList[Plus, 0, {a, b, c}]
{0, a, a + b, a + b + c}

rn = FoldList[
  Plus, {0, 0}, {{0, 1}, {1, 0}, {0, -1}, {-1, 0}}[[# + 1]] &
  /@ walk]

{{0, 0}, {1, 0}, {2,
  0}, {2, -1}, {2, -2}, {2, -3}, {2, -2}, {2, -1}, {2, -2},
 {3, -2}, {2, -2}, {2, -3}, {3, -3}, {3, -4}, {4, -4}, {4, -5},
 {5, -5}, {4, -5}, {3, -5}, {2, -5}, {2, -4}, {1, -4}, {1, -5},
 {0, -5}, {-1, -5}, {-1, -4}, {-2, -4}, {-2, -3}, {-2, -4},
 {-1, -4}, {-1, -3}, {-1, -2}, {-1, -3}, {-1, -2}}
```

   All we have to do now is to connect these points together to get a random walk.

---
[1] http://mathgis.blogspot.com/

```
Graphics[{Line[rn], PointSize[Large], Green, Point[First@rn],
Red, Point[Last@rn]}]
```

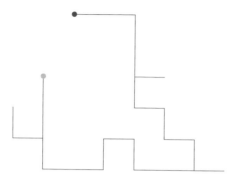

Figure 14.15 Random walks for $\sqrt{2}$

Putting all these codes together, and starting with 10000 decimal digits of $\sqrt{2}$, we have

```
x = N[Sqrt[2], 10000];
walk = First@RealDigits[x, 4];
rn = FoldList[
   Plus, {0, 0}, {{0, 1}, {1, 0}, {0, -1}, {-1, 0}}[[# + 1]] &
   /@ walk];
Graphics[{Line[rn], PointSize[Large], Green, Point[First@rn],
Red, Point[Last@rn]}]
```

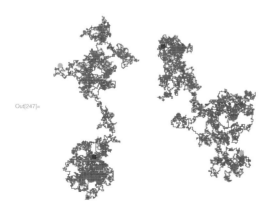

Figure 14.16 Random walks for $\sqrt{2}$ and $\sqrt{3}$

If one wants to see how the random walk actually develops, one can wrap
Manipulate around the whole code, and let the value of n run from 1 to 10000
and see the fabulous result.

```
Manipulate[x = N[Sqrt[13], n];
walk = First@RealDigits[x, 4];
rn = FoldList[
   Plus, {0, 0}, {{0, 1}, {1, 0}, {0, -1}, {-1, 0}}[[# + 1]] &
   /@ walk];
Graphics[{Line[rn], PointSize[Large], Green, Point[First@rn],
Red, Point[Last@rn]}], {n, 1, 10000, 1}]
```

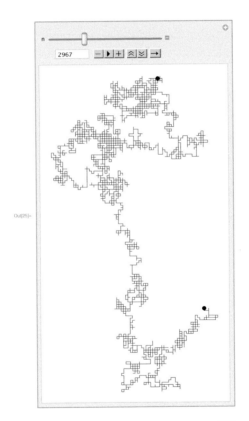

Figure 14.17 Random walk for $\sqrt{13}$

—— Problem 14.9

Define the Conway recursive sequence $a(1) = 1, a(2) = 1$ and

$$a(n) = a(a(n-1)) + a(n - a(n-1))$$

in *Mathematica*, and plot $a(n)/n$, as n runs from 1 to 1500.

$\Longrightarrow$ SOLUTION.

Calculating $a(n)/n$ for $1 \leq n \leq 1500$ gives us 1500 numbers. Using ListPlot we can put these numbers into a figure. Recall from Chapter 12 how we handle recursive functions in *Mathematica*:

```
a[1] = a[2] = 1
a[n_] := a[n] = a[a[n - 1]] + a[n - a[n - 1]]
```

Here is the value of $a(n)/n$ for the first 20 numbers in the Conway sequence:

```
t = (a /@ Range[20])/Range[20]

{1, 1/2, 2/3, 1/2, 3/5, 2/3, 4/7, 1/2, 5/9, 3/5, 7/11,
7/12, 8/13, 4/7, 8/15, 1/2, 9/17, 5/9, 11/19, 3/5}

t = (a /@ Range[1500])/Range[1500];

?ListPlot
LisPlot[{y1,y2,...}] plots points corresponding to a list of
values, assumed to correspond to x coordinates 1, 2, ...

LisPlot[{x1,y1},{x2,y2},...}] plots a list of points with
specified x and y coordinates.

ListPlot[t]
```

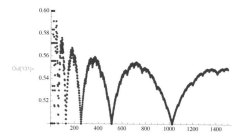

Figure 14.18 The Conway sequence

It is interesting to see that a slight change in the definition of the Conway recursive function makes the behaviour of the sequence quite chaotic. Let us define a very similar recursive function to the Conway one by $b(1) = b(2) = 1$ and[2]

$$b(n) = b(b(n-1)) + b(n-1-b(n-2)).$$

Creating a similar graph as above for this function we have

```
t1 = (b /@ Range[1500])/Range[1500];

GraphicsGrid[{{ListPlot[t], ListPlot[t1]}}]
```

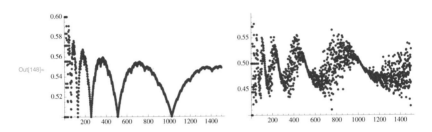

Figure 14.19 The Conway sequence and its cousin

Notice how we have used `GraphicsGrid` to put two plots in one row (see Fig. 14.19).

⎿_____

_____ Problem 14.10

Recall from Problem 9.2 that to get the Thue–Morse sequence, one starts with 0 and then repeatedly replaces 0 with 01 and 1 with 10. Create the 15th number in the Thue–Morse sequence. Then, based on this number, create a graph as follows: starting from $(0,0)$, move ahead by one unit if you encounter 1 in the Thue–Morse sequence and rotate counterclockwise by an angle of $-\pi/3$ if you encounter 0. (The resulting curve converges to the Koch snowflake, a fractal curve of infinite length containing a finite area.[3])

⟹ SOLUTION.

We first create the 15th Thue–Morse number. See Problem 9.2, where we have written this code. Note that this is a huge number and it might take some

---

[2] This has been studied by K. Pinn, A chaotic cousin of Conway's recursive sequence, available at arXiv.org.

[3] See, for example, the Thue–Morse sequence in Wikipedia.

time until *Mathematica* produces this, and we are not going to print the output. To create the graph, we will use a similar technique to that in Problem 14.8.

```
Flatten[ReplaceRepeated[{0}, {0 -> {0, 1}, 1 -> {1, 0}},
    MaxIterations -> 4]]
```

ReplaceRepeated::rrlim: Exiting after {0} scanned 4 times. >>

{0, 1, 1, 0, 1, 0, 0, 1, 1, 0, 0, 1, 0, 1, 1, 0}

```
Flatten[ReplaceRepeated[{0}, {0 -> {0, 1}, 1 -> {1, 0}},
    MaxIterations -> 4]] /. {1 -> -Pi/3}
```

ReplaceRepeated::rrlim: Exiting after {0} scanned 4 times. >>

{0, -Pi/3, -Pi/3, 0, -Pi/3, 0, 0, -Pi/3, -Pi/3, 0, 0,
-Pi/3, 0, -Pi/3, -Pi/3, 0}

```
Accumulate[
 Flatten[ReplaceRepeated[{0}, {0 -> {0, 1}, 1 -> {1, 0}},
    MaxIterations -> 4]] /. {1 -> -Pi/3}]
```

ReplaceRepeated::rrlim: Exiting after {0} scanned 4 times. >>

{0, -Pi/3, -2 Pi/3, -2 Pi/3, -Pi, -Pi, -Pi, -4Pi/3, -5Pi/3),
-5Pi/3, -5Pi/3, -2Pi, -2Pi, -7Pi/3, -8Pi/3, -8Pi/3}

```
re = {Sin[#], Cos[#]} & /@
    Accumulate[
     Flatten[ReplaceRepeated[{0}, {0 -> {0, 1}, 1 -> {1, 0}},
        MaxIterations -> 15]] /. {1 -> -Pi/3}]
```

$$\left\{ \{0,1\}, \left\{-\tfrac{\sqrt{3}}{2}, \tfrac{1}{2}\right\}, \left\{-\tfrac{\sqrt{3}}{2}, -\tfrac{1}{2}\right\}, \left\{-\tfrac{\sqrt{3}}{2}, -\tfrac{1}{2}\right\}, \{0,-1\}, \{0,-1\}, \right.$$
$$\{0,-1\}, \left\{\tfrac{\sqrt{3}}{2}, -\tfrac{1}{2}\right\}, \left\{\tfrac{\sqrt{3}}{2}, \tfrac{1}{2}\right\}, \left\{\tfrac{\sqrt{3}}{2}, \tfrac{1}{2}\right\}, \left\{\tfrac{\sqrt{3}}{2}, \tfrac{1}{2}\right\}, \{0,1\}, \{0,1\},$$
$$\left\{-\tfrac{\sqrt{3}}{2}, \tfrac{1}{2}\right\}, \left\{-\tfrac{\sqrt{3}}{2}, -\tfrac{1}{2}\right\}, \left\{-\tfrac{\sqrt{3}}{2}, -\tfrac{1}{2}\right\}, \{0,-1\}, \{0,-1\}, \{0,-1\},$$
$$\cdots\ 32\,731\ \cdots, \left\{-\tfrac{\sqrt{3}}{2}, -\tfrac{1}{2}\right\}, \left\{-\tfrac{\sqrt{3}}{2}, -\tfrac{1}{2}\right\}, \{0,-1\}, \{0,-1\}, \{0,-1\},$$
$$\left\{\tfrac{\sqrt{3}}{2}, -\tfrac{1}{2}\right\}, \left\{\tfrac{\sqrt{3}}{2}, -\tfrac{1}{2}\right\}, \left\{\tfrac{\sqrt{3}}{2}, \tfrac{1}{2}\right\}, \{0,1\}, \{0,1\}, \{0,1\}, \left\{-\tfrac{\sqrt{3}}{2}, \tfrac{1}{2}\right\},$$
$$\left\{-\tfrac{\sqrt{3}}{2}, -\tfrac{1}{2}\right\}, \left\{-\tfrac{\sqrt{3}}{2}, -\tfrac{1}{2}\right\}, \{0,-1\}, \{0,-1\}, \{0,-1\}, \left\{\tfrac{\sqrt{3}}{2}, -\tfrac{1}{2}\right\}\right\}$$

| large output | show less | show more | show all | set size limit... |

ReplaceRepeated::rrlim: Exiting after {0} scanned 15 times. >>

```
re1 = FoldList[Plus, {0, 0}, re];
```

```
Graphics[Line[re1]]
```

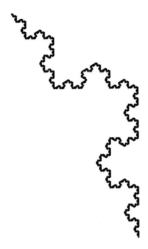

Figure 14.20 Thue–Morse fractal

14.2 Three-dimensional graphs

As in Section 14.1 for two-dimensional graphics, *Mathematica* provides several commands to handle three-dimensional graphics. A table similar to the one on page 205 can be drawn to show which command should be used in different circumstances.

Function	Example	Graphics command
$z = f(x, y)$	$z = \sin(x^2 + y^2)e^{-x^2}$	Plot3D
$x = f(t), y = g(t),$	$x = \sin(3t), y = \cos(4t),$	ParametricPlot3D
$z = h(t)$	$z = \sin(5t)$	
$f(x, y, z) = 0$	$6x^2 - 2x^4 - y^2z^2 = 0$	ContourPlot3D
$f(x, y, z) \geq 0$	$x^4 + (x - 2y^2) \geq 0$	RegionPlot3D

Problem 14.11

Plot the graph of the "cowboy hat" equation

$$\sin(x^2 + y^2)e^{-x^2} + \cos(x^2 + y^2)$$

as both x and y range from -2 to 2.

$\Longrightarrow$ SOLUTION.

We first translate the equation into *Mathematica* and then, using `Plot3D`, we plot the graph.

```
q[x_,y_]:=Sin[x^2 + y^2] Exp[-x^2] + Cos[x^2 + y^2]

Plot3D[q[x, y], {x, -2, 2}, {y, -2, 2}, PlotPoints -> 50]
```

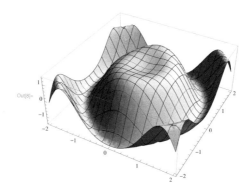

L_____

_____ Problem 14.12

Plot the graph of $f(x,y) = xy\sin(x^2)\cos(y^2)$ when $-2\pi \le x \le 0$ and $-2\pi \le y \le 0$. Using `PlotPoints` plot the graph with a different accuracy.

$\Longrightarrow$ SOLUTION.

Here we use `PlotPoints` to force *Mathematica* to generate less or more sample points than it usually does. One can see, if we ask *Mathematica* to produce the graph based on only 5 sample points, we get a crude sort of a graph. Changing the `PlotPoints` to 50, we will get a high quality graph.

```
Plot3D[x y Sin[x^2] Cos[y^2], {x, -2 Pi, 0}, {y, -2 Pi, 0},
    PlotRange -> All, PlotPoints -> 5]

Plot3D[x y Sin[x^2] Cos[y^2], {x, -2 Pi, 0}, {y, -2 Pi, 0},
    PlotRange -> All, PlotPoints -> 50]
```

One nice experiment is to observe how a graph becomes more accurate as the number of `PlotPoints` increases. One way to do this is to use `Manipulate`,

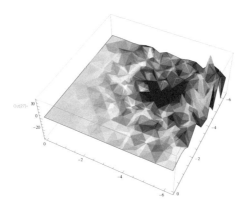

Figure 14.21 $xy \sin(x^2) \cos(y^2)$ with 5 sample points

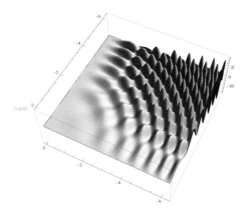

Figure 14.22 $xy \sin(x^2) \cos(y^2)$ with 50 sample points

and let `PlotPoints->i` and change i from, say, 5 to 30. However, as it takes a long time for each of these graphs to be produced, one does not get a smooth animation. One solution is to produce several instances and show them one after the other using `ListAnimate`. The following code does just that: we produce several "frames" of the graph, using `Table`, which changes the `PlotPoints`, and then show the frames one after the other using `ListAnimate`.

```
ListAnimate[
  Table[Plot3D[x y Sin[x^2] Cos[y^2], {x, -2 Pi,0},{y, -2 Pi,0},
    PlotRange -> All, PlotPoints -> i], {i, 5, 30, 3}]]
```

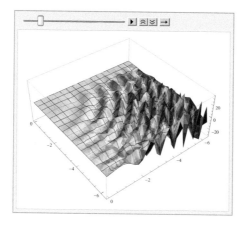

Figure 14.23 $xy \sin(x^2) \cos(y^2)$ using `ListAnimate`

Problem 14.13

Plot the graph of $6x^2 - 2x^4 - y^2 z^2 = 0$.

$\Longrightarrow$ SOLUTION.

```
ContourPlot3D[6 x^2 - 2 x^4 - y^2 z^2 == 0, {x, -10, 10},
{y, -10, 10}, {z, -10, 10}]
```

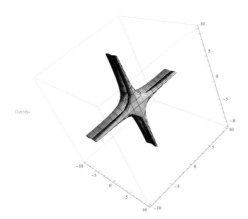

Problem 14.14

Plot the graph of $(x^2 + \frac{9}{4}y^2 + z^2 - 1)^3 - x^2 z^3 - \frac{9}{80}y^2 z^3 = 0$.

$\implies$ SOLUTION.

```
ContourPlot3D[(x^2 + 9/4 y^2 + z^2 - 1)^3 - x^2 z^3 - 9/80 y^2 z^3 ==
  0, {x, -2, 2}, {y, -2, 2}, {z, -2, 2}]
```

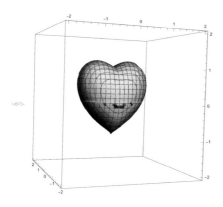

Exercise 14.10

Plot the graphs of the following.[4]

Calyx	$x^2 + y^2 z^3 = z^4$
Durchblick	$x^3 y + x z^3 + y^3 z + z^3 + 5z = 0$
Seepferdchen	$(x^2 - y^3)^2 = (x + y^2)z^3$
Geisha	$x^2 yz + x^2 z^2 = y^3 z + y^3$
Schneeflocke	$x^3 + y^2 z^3 + yz^4 = 0$

---

[4] A nice gallery of these and much more can be found on Herwig Hauser's homepage
https://homepage.univie.ac.at/herwig.hauser/gallery.html

_____ Problem 14.15

Plot the graph of $x = \sin(3t), y = \cos(4t), z = \sin(5t)$, for $-\pi \le t \le \pi$. Then create a dynamic setting and plot the graph of $x = \sin(nt), y = \cos(mt), z = \sin(5t)$ for $1 \le n, m \le 10$.

$\Longrightarrow$ SOLUTION.

```
ParametricPlot3D[{Sin[3 t], Cos[4 t], Sin[5 t]}, {t, -Pi, Pi},
   PlotStyle -> Tube[0.05]]
```

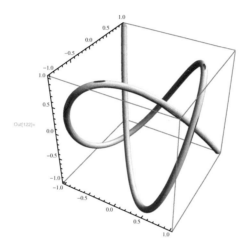

```
Manipulate[
 ParametricPlot3D[{Sin[n t], Cos[m t], Sin[5 t]}, {t, -Pi, Pi},
   PlotStyle -> Tube[0.05]], {n, 1, 10, 1}, {m, 1, 10, 1}]
```

_____ Problem 14.16

Consider the function $f(x, y) = \frac{\sin(x^2+y^2)}{x+y}$. Plot the contour of

$$f\Big(f\big(f(x, y), f(x, y)\big), f\big(f(x, y), f(x, y)\big)\Big).$$

$\Longrightarrow$ SOLUTION.

The command to be used is ContourPlot. We define the function f[x,y] and translate the expression into *Mathematica*.

```
f[x_, y_] := Sin[x^2 + y^2]/(x + y)

ContourPlot[
 f[f[f[x, y], f[x, y]], f[f[x, y], f[x, y]]],
   {x, -Pi, Pi}, {y, -Pi, Pi}]
```

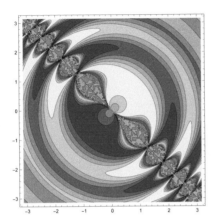

14.3 Plotting data

It is very easy to present data in *Mathematica*. There are numerous functions
to plot data, among them `ListPlot` and `DateListPlot`.

In this section we demonstrate these functions by importing the data of
coffee consumptions of three countries in the last 40 years and plot the data in
an interactive manner. This data is saved in three XLS files. We first import
them by using the command `Import`, and finding the path by using File Path
from the Insert menu.

```
aItaly = Import[
    "/Users/hazrat/Desktop/coffee Italy history.xls",
      {"Data", 1}];
```

This imports the coffee consumption in Italy for the last 40 years. The
expression {"Data", 1} in the command `Import` tells *Mathematica* that the
data is of the form `Data`, as opposed to say, `Image`, and 1 uploads the first page
of the spreadsheet.

Let us look at some of the data using `Short`.

```
Short[aItaly, 10]
```

{{Date, 1000.}, { [📅] Sun 1 Jan 1961 00:00:00 GMT+10. , 105.3},

{ [📅] Mon 1 Jan 1962 00:00:00 GMT+10. , 111.9},

{ [📅] Tue 1 Jan 1963 00:00:00 GMT+10. , 116.8}, { [📅] Wed 1 Jan 1964 00:00:00 GMT+10. , 119.3},

{ [📅] Fri 1 Jan 1965 00:00:00 GMT+10. , 120.2}, { [📅] Sat 1 Jan 1966 00:00:00 GMT+10. , 123.4},

{ [📅] Sun 1 Jan 1967 00:00:00 GMT+10. , 145.2}, ≪32≫,

{ [📅] Sat 1 Jan 2000 00:00:00 GMT+10. , 307.1}, { [📅] Mon 1 Jan 2001 00:00:00 GMT+10. , 312.3},

{ [📅] Tue 1 Jan 2002 00:00:00 GMT+10. , 306.}, { [📅] Wed 1 Jan 2003 00:00:00 GMT+10. , 326.7},

{ [📅] Thu 1 Jan 2004 00:00:00 GMT+10. , 321.}, { [📅] Sat 1 Jan 2005 00:00:00 GMT+10. , 335.1},

{ [📅] Sun 1 Jan 2006 00:00:00 GMT+10. , 337.8}, { [📅] Mon 1 Jan 2007 00:00:00 GMT+10. , 342.8}}

We remove the first entry and then plot the data.

```
aItaly = Rest[aItaly];
```

```
DateListPlot[aItaly]
```

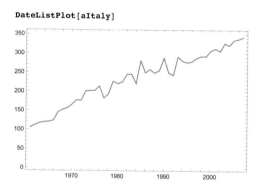

We proceed by importing the coffee consumption for Australia.

```
aAustralia =
  Import["/Users/hazrat/Desktop/coffee Australia history.xls", {"Data", 1}];
Short[aAustralia, 10]
```

{{Date, 1.}, { [📅] Sun 1 Jan 1961 00:00:00 GMT+10. , 22 199.},

{ [📅] Mon 1 Jan 1962 00:00:00 GMT+10. , 18 948.},

{ [📅] Tue 1 Jan 1963 00:00:00 GMT+10. , 13 876.},

{ [📅] Wed 1 Jan 1964 00:00:00 GMT+10. , 16 150.}, { [📅] Fri 1 Jan 1965 00:00:00 GMT+10. , 16 254.},

{ [📅] Sat 1 Jan 1966 00:00:00 GMT+10. , 13 957.}, { [📅] Sun 1 Jan 1967 00:00:00 GMT+10. , 15 950.},

≪32≫, { [📅] Sat 1 Jan 2000 00:00:00 GMT+10. , 54 272.},

{ [📅] Mon 1 Jan 2001 00:00:00 GMT+10. , 40 508.},

As with the data for Italy, we need to drop the first entry. Note also that here the data is given per tons, so we need to divide them by 1000 to be in line with the data from Italy which was given per 1000 tons.

```
aAustralia = Rest[aAustralia];
aAustralia = aAustralia /. {x_, y_} -> {x, y/1000};
```

Next we import the consumption in England.

```
aEngland =
  Import["/Users/hazrat/Desktop/coffee England history.xls",
  {"Data", 1}];

aEngland = Rest[aEngland];
```

Here is the plot of data of all three countries. As is clear from `PlotStyle`, England is shown by red, Australia by green and Italy is shown by blue.

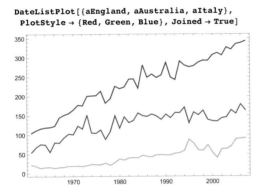

Finally, we are going to represent the data in an interactive manner using the mighty `Manipulate`.

Recall the command ;; from Chapter 4 on how to access a sequence of elements of a list. The following code shows how to present the data:

```
aItaly[[1]]
```

{ 🗓 Sun 1 Jan 1961 00:00:00 GMT+10. , 105.3}

```
aItaly[[{1, 2}]]
```

{{ 🗓 Sun 1 Jan 1961 00:00:00 GMT+10. , 105.3}, { 🗓 Mon 1 Jan 1962 00:00:00 GMT+10. , 111.9}}

```
aItaly[[1 ;; 3]]
```

{{ 🗓 Sun 1 Jan 1961 00:00:00 GMT+10. , 105.3},

{ 🗓 Mon 1 Jan 1962 00:00:00 GMT+10. , 111.9}, { 🗓 Tue 1 Jan 1963 00:00:00 GMT+10. , 116.8}}

We use this with a `Manipulate`.

```
Manipulate[DateListPlot[aItaly[[1 ;; i]], Joined → True], {i, 1, 47, 1}]
```

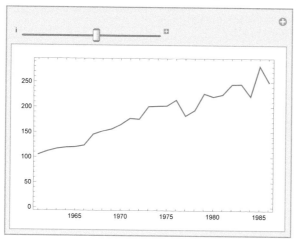

Next, we are going to represent the consumption of all three countries.

```
Manipulate[DateListPlot[{aEngland[[1 ;; i]],
  aItaly[[1 ;; i]], aAustralia[[1 ;; i]]},
  Joined → True, PlotStyle → {Red, Blue, Green},
  PlotRange → {{{1960}, {2007}}, {10, 400}}], {i, 1, 47, 1}]
```

Finally, a bit more decoration on the output.

```
Manipulate[DateListPlot[{aEngland[[1 ;; i]],
   aItaly[[1 ;; j]], aAustralia[[1 ;; k]]}, Joined → True,
  PlotStyle → {Red, Blue, Green}, PlotRange → {{{1960}, {2010}}, {10, 400}}],
 {{i, 1, Text[Style[England, 14, Red]]}, 1, 47, 1},
 {{j, 1, Text[Style[Italy, 14, Blue]]}, 1, 47, 1},
 Delimiter,
 {{k, 1, Text[Style[Australia, 14, Green]]}, 1, 47, 1}]
```

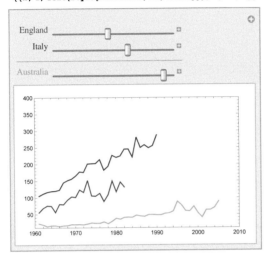

15
Calculus and equations

Mathematica comes with several powerful commands to solve different kinds of equations. Doing calculus is also one of the strong features of this software. This chapter gives a brief account of what is available here.

15.1 Solving equations

Solving equations and finding roots for different types of equations and relations are one of the main endeavours of mathematics. For polynomials with one variable, i.e., of the form $a_n x^n + a_{n-1} x^{n-1} + \cdots + a_1 x + a_0$, it has been proved that there is no formula for finding the roots when $n \geq 5$ (in fact, when $n = 3$ or 4, the formulas are not that pretty!). This forces us to find numerical ways to estimate the roots of the equations. Using Wolfram *Mathematica*® we have several commands at our disposal. There are different kinds of equations and they require different commands to find the roots. The following table shows which command is suitable for different formats of equations.

Example	Commands to solve an equation
$x^4 - 3x^3 + 2x + 10 = 0$	Solve
$x^6 - 4x^3 + 12x + 10 = 0$	NSolve
$x^3 - 3x^2 + 5 < 0$	Reduce
$x^{x-10} = x^2$ and $x \in \mathbb{N}$	FindInstance
$\sin(x) = x - 1$	FindRoot

Although all the examples in the above table are equations with one variable, *Mathematica* can also handle equations with more than one variable. The

© Springer International Publishing Switzerland 2015
R. Hazrat, *Mathematica®: A Problem-Centered Approach*, Springer Undergraduate Mathematics Series, DOI 10.1007/978-3-319-27585-7_15

point is, in order to find roots for the equations, one needs to experiment with the above commands to find which one produces the solutions to the equation. In the problems below we consider several situations to demonstrate how one works with these commands.

▬▬ Problem 15.1

Find the roots of the following equations:

$$x^4 - 3x^3 + 2x + 10 = 0,$$
$$x^6 - 4x^3 + 12x + 10 = 0,$$
$$x^3 - 3x^2 + 5 < 0,$$
$$x^{x-10} = x^2 \text{ and } x \in \mathbb{N},$$
$$\sin(x) = x - 1.$$

⟹ SOLUTION.
To solve the first equation, one can use `Solve`, as this is a polynomial of degree 4, thus there is an algebraic method to get all the solutions. We then use `N` to get the numerical value of these solutions.

```
N[Solve[x^4 - 3 x^3 + 2 x + 10 == 0, x]]
```

```
{{x -> -0.822108 - 1.00134 I}, {x -> -0.822108 +
    1.00134 I}, {x -> 2.32211 - 0.75191 I}, {x ->
    2.32211 + 0.75191 I}}
```

For the second equation, `Solve` is not going to give us any answer. However, using `NSolve`, we get the following roots:

```
NSolve[x^6 - 4 x^3 + 12 x + 10 == 0, x]
```

```
{{x -> -0.929435 - 0.361625 I}, {x -> -0.929435 +
    0.361625 I}, {x -> -0.62568 - 1.72428 I},
    {x -> -0.62568 + 1.72428 I}, {x -> 1.55512 - 0.754856 I},
    {x -> 1.55512 + 0.754856 I}}
```

For the third inequality, `Reduce` proved to be the right tool:

```
Reduce[x^3 - 3 x^2 + 5 < 0, x]
x < Root[5 - 3 #1^2 + #1^3 &, 1]
```

```
N[%]
x < -1.1038
```

For the fourth equation, we have:

```
FindInstance[x^(x - 10) == x^2, x, Integers]
{{x -> 12}}
```

In order to find solutions for the equation $\sin(x) = x - 1$, as this is not an algebraic equation, there is no hope of being able to get any meaningful answer using any of the commands above, as the following shows:

```
Solve[Sin[x] == x - 1, x]
```

During evaluation of In[200]:= Solve::tdep: The equations appear to involve the variables to be solved for in an essentially non-algebraic way. >>

One gets the same message using NSolve and Reduce. The way forward is to plot the graph of this equation and see where the curve crosses the x-axis:

```
Plot[Sin[x] - x + 1, {x, -2 Pi, 2 Pi}]
```

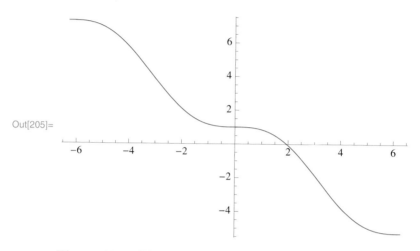

Out[205]=

Figure 15.1 Plotting the graph and using FindRoot

From the graph it is clear that this equation has a root somewhere close to 2. Now using FindRoot and helping *Mathematica* a little, that is, giving her a point close to the root, she can find the exact root for us:

```
FindRoot[Sin[x] == x - 1, {x, 2}]
{x -> 1.93456}
```

As we mentioned, all the commands in the above table can also handle equations with more than one variable.

Problem 15.2

Find all pairs of real numbers (x, y) satisfying the system of equations

$$2 - x^3 = y$$
$$2 - y^3 = x + \sin(y).$$

⟹ SOLUTION.

Let us first plot the graphs of these two functions in one figure:

```
ContourPlot[{2 - x^3 == y, 2 - y^3 == x + Sin[y]}, {x, -100,
  100}, {y, -100, 100}]
```

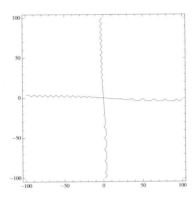

Figure 15.2 Graph of $2 - x^3 = y$ and $2 - y^3 = x + \sin(y)$

The figure shows that these two graphs intersect each other in just one point. Let us concentrate on a smaller interval:

```
ContourPlot[{2 - x^3 == y, 2 - y^3 == x + Sin[y]}, {x, -10,
  10}, {y, -10, 10}]
```

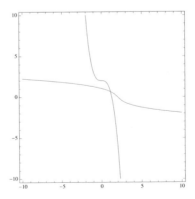

Figure 15.3 Plotting the graphs and using FindRoot

So we can easily find this point by using FindRoot and giving a point close to this root:

```
FindRoot[{2 - x^3 == y, 2 - y^3 == x + Sin[y]}, {x, 2}, {y, 1}]
{x -> 1.10294, y -> 0.658304}
```

Problem 15.3

Find the solutions of the equation $\sin\left(x^2\right) - \cos\left(x^3\right) = 0$, for $0 \le x \le \pi$.

$\Longrightarrow$ SOLUTION.

One can try to find roots of this equation by NSolve or Reduce. However it will not take long to realise that this is not the right approach. We first plot the graph of this equation for $0 \le x \le \pi$.

```
Plot[Sin[x^2] - Cos[x^3], {x, 0, Pi}]
```

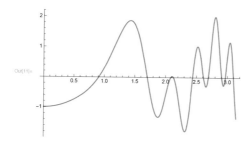

The graph around 2 seems to hit the x-axis in several places. We plot the graph around this region to get a better view.

```
Plot[Sin[x^2] - Cos[x^3], {x, 2, 2.3}]
```

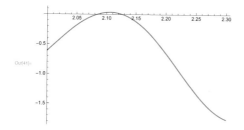

Now we choose values close to the roots (where the graph intersects with the x-axis) and then use FindRoot to find the roots. We collects these point in a list and represent them in the graph along with the equation using Epilog. Note how we use the rules and pattern matching to change the values of the points. This was studied in detail in Problem 10.2.

```
s = {1, 1.65, 2, 2.15, 2.4, 2.6, 2.7, 2.9, 2.98, 3.1};

t = s /. n_?NumberQ -> Point[{n, 0}]

{Point[{1, 0}], Point[{1.65, 0}], Point[{2, 0}],
 Point[{2.15, 0}], Point[{2.4, 0}], Point[{2.6, 0}],
 Point[{2.7, 0}], Point[{2.9, 0}], Point[{2.98, 0}],
 Point[{3.1, 0}]}

Plot[Sin[x^2] - Cos[x^3], {x, 0, Pi}, Epilog -> t]
```

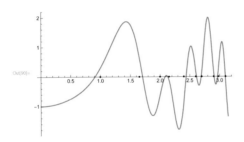

Next we define FindRoot as a pure function and we send it into the list to get all the solutions in one go.

```
s1 = FindRoot[Sin[x^2] - Cos[x^3], {x, #}] & /@ s

{{x -> 0.907468}, {x -> 1.70421}, {x -> 2.08451},
 {x -> 2.12645}, {x -> 2.43736}, {x -> 2.61184},
 {x -> 2.69009}, {x -> 2.90604}, {x -> 2.9652},
 {x -> 3.09625}}

t1 = Flatten[s1 /. (x -> n_) -> Point[{n, 0}]]

{Point[{0.907468, 0}], Point[{1.70421, 0}],
 Point[{2.08451, 0}], Point[{2.12645, 0}],
 Point[{2.43736, 0}], Point[{2.61184, 0}],
 Point[{2.69009, 0}], Point[{2.90604, 0}],
 Point[{2.9652, 0}], Point[{3.09625, 0}]}

Plot[Sin[x^2] - Cos[x^3], {x, 0, Pi}, Epilog -> t1]
```

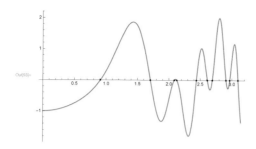

Out[93]=

▬ Problem 15.4

Find two positive integers x and y such that the equation $f(x,y) = 2xy^4 + x^2y^3 - 2x^3y^2 - y^5 - x^4y + 2y$ gives the 10th Fibonacci number. (For your information: There is an extraordinary theorem which states that the Fibonacci numbers are precisely the positive values of this equation where x and y are integers.)

⟹ SOLUTION.

As we are looking for a sample solution of this equation, FindInstance is the right command:

```
f[x_, y_] := 2 x y^4 + x^2 y^3 - 2 x^3 y^2 - y^5 - x^4 y + 2 y
```

```
FindInstance[f[x, y] == Fibonacci[10], {x, y}, Integers]
{{x -> -89, y -> 55}}
```

The x and y we got are not positive. We try to find more sample solutions:

```
FindInstance[f[x, y] == Fibonacci[10], {x, y}, Integers, 2]
{{x -> -89, y -> 55}, {x -> 34, y -> 55}}
```

▬ Problem 15.5

1. Consider the equations $x^2y + xy^2 = 2$ and $x^3y + xy^3 = 3$. Find two pairs (x,y) of real values which satisfy these equations.

2. There are positive real numbers x and y such that $x^x + y^y = 2$, for example $x = y = 1$. Plot a graph showing the solutions (x,y) of this equation for $x, y \geq 0$.

$\implies$ SOLUTION.

We plot these equations in one go with the help of `ContourPlot`.

```
g = ContourPlot[{x^2 y + x y^2 == 2, x^3 y + x y^3 == 3},
{x, 0, 10}, {y, 0, 10}, PlotTheme -> "Scientific"]
```

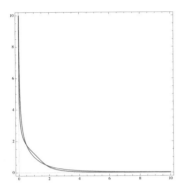

The graphs clearly intersect in two points. We use `FindRoot` to find the roots. We give *Mathematica* the (x, y) coordinate close to the the intersections.

```
p1 = FindRoot[{x^2 y + x y^2 == 2, x^3 y + x y^3 == 3},
{{x, 2}, {y, 0.5}}]
```

```
{x -> 1.77939, y -> 0.494336}
```

```
p2 = FindRoot[{x^2 y + x y^2 == 2, x^3 y + x y^3 == 3},
{{x, 0.5}, {y, 2}}]
```

```
{x -> 0.494336, y -> 1.77939}
```

We retrieve these solutions and then using `Point` represent them in the same graph.

```
{x, y} /. {p1, p2}
{{1.77939, 0.494336}, {0.494336, 1.77939}}
```

```
x1 = Graphics[{PointSize[Large], Point[{x, y} /. {p1, p2}]}];
```

The second part of the problem is similar. We will plot the graph, showing the solutions.

```
ContourPlot[{x^y + y^x == 2, x - y == 0}, {x, 0, 10}, {y, 0, 10}]
```

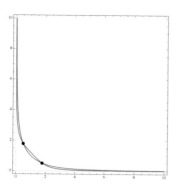

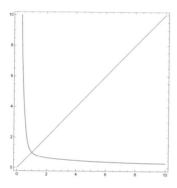

Problem 15.6

Let A_n be the $n \times n$ matrix with (i, j)-th entry equal to $x^{|i-j-ij|+ij}$. For $n = 3$ and 4 find exactly all values of x for which the determinant of A_n is zero. Find all the distinct values of x for which the determinant of A_6 is zero. Observe that there are 21 of them.

$\implies$ SOLUTION.

The only challenge here is to define this matrix. For this see Chapter 13. This done, we start with Solve to see whether we get exact roots of the equation produced by the determinant of A_3.

```
A[n_] := Array[x^(Abs[#1 - #2 - #1 #2] + #1 #2) &, {n, n}]

Det[A[3]]
-x^20 + 2 x^22 - 2 x^26 + x^28
```

```
Solve[Det[A[3]] == 0]
{{x -> -1}, {x -> -1}, {x -> -1}, {x -> 0}, {x -> 0}, {x ->
    0}, {x -> 0}, {x -> 0}, {x -> 0}, {x -> 0}, {x -> 0}, {x ->
    0}, {x -> 0}, {x -> 0}, {x -> 0}, {x -> 0}, {x -> 0}, {x ->
    0}, {x -> 0}, {x -> 0}, {x -> 0}, {x -> 0}, {x ->
    0}, {x -> -I}, {x -> I}, {x -> 1}, {x -> 1}, {x -> 1}}
```

This shows the command Solve is successful in finding the roots. So we stick to it. Also the above result shows there are many repeated roots. As we are interested in distinct roots, we get rid of the repetition by using Union.

```
Union[Solve[Det[A[4]] == 0]]
{{x -> -1}, {x -> 0}, {x -> -I}, {x -> I}, {x ->
    1}, {x -> -(-1)^(1/3)}, {x -> (-1)^(
    1/3)}, {x -> -(-1)^(2/3)}, {x -> (-1)^(2/3)}}
```

For the last part of the problem:

```
Length[Union[Solve[Det[A[6]] == 0]]]
21
```

Problem 15.7

Consider the system of equations

$$\begin{cases} 3x + 2y - 4z = 2 \\ 3x + 2y - (4+R)z = \frac{1}{2} \\ 3x + (2+S)y - 4z = 3 \end{cases} \tag{15.1}$$

First find the unique solution for this system when $R = 0.05$ and $S = -0.05$. Next explore if R and S change, how the solution behaves, and how far the new solution would be from initial solution.

$\Longrightarrow$ SOLUTION.

This interesting problem was explored by Shiskowski and Frinkle (Example 3.3.5).[1] We will follow them here to explore the behaviour of solutions with respect to R and S.

First for $R = 0.05$ and $S = -0.05$, Equation 15.1 takes the form

```
{3 x + 2 y - 4 z == 2, 3 x + 2 y - (4 + R) z == 1/2,
    3 x + (2 + S) y - 4 z == 3} /. {R -> 0.05, S -> -0.05}
```

```
{3 x + 2 y - 4 z == 2, 3 x + 2 y - 4.05 z == 1/2,
    3 x + 1.95 y - 4 z == 3}
```

---

[1] K. Shiskowski, K. Frinkle, Principles of Linear Algebra with Mathematica, Wiley & Sons, 2011.

Thus solving this system we get the unique solution

```
Solve[{3 x + 2 y - 4 z == 2, 3 x + 2 y - (4 + R) z == 1/2,
    3 x + (2 + S) y - 4 z == 3} /. {R -> 0.05, S -> -0.05},
   {x, y, z}]
```

```
{{x -> 54., y -> -20., z -> 30.}}
```

If we plot these planes for $R = 0.05$ and $S = -0.05$, it would be clear why the solution is very sensitive to changes of R and S. The planes are nearly parallel and thus slight changes to R or S would change the points at which these planes intersect substantially.

```
ContourPlot3D[{3 x + 2 y - 4 z = 2, 3 x + 2 y - 4.05 z = 1/2, 3 x + 1.95 y - 4 z = 3},
   {x, -6, 6}, {y, -6, 6}, {z, -6, 6}, PlotTheme → "Detailed"]
```

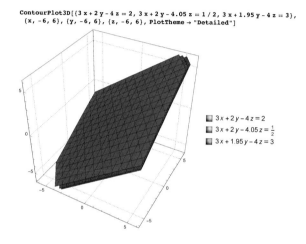

Next, let us look at the general solution of Equation 15.1:

```
Solve[{3 x + 2 y - 4 z == 2, 3 x + 2 y - (4 + R) z == 1/2,
    3 x + (2 + S) y - 4 z == 3}, {x, y, z}]
```

```
{{x -> (2 (-R + 3 S + R S))/(3 R S), y -> 1/S,
   z -> 3/(2 R)}}
```

The standard procedure to find solutions for a system of equations is to use row reductions. *Mathematica* can perform this procedure and we can get the solutions to the above equations using row reduction as well.

```
RAs = RowReduce[As]
```

```
{{1, 0, 0, (2 (-R + 3 S + R S))/(3 R S)}, {0, 1, 0, 1/S},
   {0, 0, 1, 3/(2 R)}}
```

We check that these general formulas are in line with the solutions we obtained for the specific R and S above.

```
approxS = RAs[[All, -1]]
{(2 (-R + 3 S + R S))/(3 R S), 1/S, 3/(2 R)}

exactS=approxS /. {R -> 0.05, S -> -0.05}
{54., -20., 30.}
```

The distance between two points (x_1, y_1, z_1) and (x_2, y_2, z_2) in space is measured by

$$\sqrt{(x_1 - y_1)^2 + (x_2 - y_2)^2 + (x_3 - y_3)^2}.$$

We define this in *Mathematica*, measuring the distance between the solution of the system of equations for $R = 0.05$ and $S = -0.05$ and other values for R and S. Clearly this function depends on R and S.

```
errorF = Sqrt[(exactS[[1]] - approxS[[1]])^2 + (exactS[[2]] -
    approxS[[2]])^2 + (exactS[[3]] - approxS[[3]])^2];
```

We are now in a position to plot the function `errorF`.

```
?ClippingStyle

ClippingStyle is an option for plotting functions that specifies
the style of what should be drawn when curves or surfaces
would extend beyond the plot range. >>

Plot3D[errorF, {R, -0.3, 0.3}, {S, -0.3, 0.3},
    ClippingStyle -> None]

ContourPlot[errorF, {R, 0.04, 0.06}, {S, -0.06, -0.04},
    Contours -> {0, 1, 2, 3, 4, 5, 6, 7, 8, 9, 10}]
```

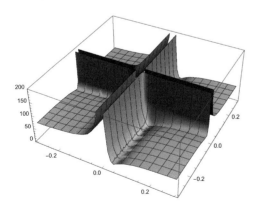

This clearly shows that when R or S tend to 0 the distance between the initial solution and the current solution increases substantially.

$\llcorner$ _____

♣ TIPS

– Using NSolve, one can ask *Mathematica* for more precision when producing the answer. NSolve[equations,var,n] would solve the equations numerically to **n** significant digits.

– Solve produces the answers as rules, whereas Reduce gives them as a Boolean statement. In addition Reduce will describe *all* possible solutions.

```
Solve[a x + b == 0, x]
{{x -> -(b/a)}}

Reduce[a x + b == 0, x]
(b == 0 && a == 0) || (a != 0 && x == -(b/a))
```

– FindRoot[Equation==0,x,x0] uses Newton's method to solve the equation. This means, if the derivative of the Equation can't be computed, Newton's method fails and so this command would not give any answer.

– FindRoot[Equation==0,x,x0,x1] uses the secant method to solve the equation. Thus, if Newton's method is not successful, one can try this approach.

Exercise 15.1

Consider the function $f(x) = x^3 - 3x + 1$. The equation $f(f(x)) = 0$ has a total of 7 distinct real roots. Find these roots.

Exercise 15.2

Find all the solutions of the equation

$$x = (1 - \frac{1}{x})^{\frac{1}{2}} + (1 + \frac{1}{x})^{\frac{1}{2}}.$$

15.2 Calculus

Two important machineries in calculus are differentiation and integration, and both use the concept of limit. We assume the reader is familiar with calculus. In the problems below we showcase some of the abilities of *Mathematica* in this area.

We start with the limit of the function

$$\lim_{n \to 0} \frac{\cos(x) - 1}{\sin(x)}.$$

Mathematica provides the command Limit for exploring the limit of a function.

```
Limit[(Cos[x] - 1)/Sin[x], x -> 0]
0
```

This is in line with the graph of the function around 0.

```
Plot[(Cos[x] - 1)/Sin[x], {x, -4, 4}]
```

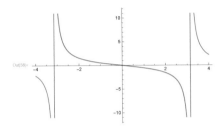

One always needs to be careful with the results of computations. Let us ask *Mathematica* to calculate $\lim_{n \to 0} 2^{-\frac{1}{x}} - \cos(x^2)$.

```
Limit[2^(-1/x) - Cos[x^2], x -> 0]
-1
```

This means the function should be continuous at 0. Let us also compute the right and left limit at this point.

```
Limit[2^(-1/x) - Cos[x^2], x -> 0, Direction -> 1]
Infinity
```

```
Limit[2^(-1/x) - Cos[x^2], x -> 0, Direction -> -1]
-1
```

This does not match the limit we calculated in the first place. Let us plot the graph of this function.

```
Plot[2^(-1/x) - Cos[x^2], {x, -2, 2}]
```

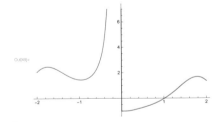

Clearly, the function is not continuous at 0. At the point 0 *Mathematica* is not correctly representing the graph. We will exclude this point from the graph to get the correct plot.

```
Plot[2^(-1/x) - Cos[x^2], {x, -2, 2}, Exclusions -> 0]
```

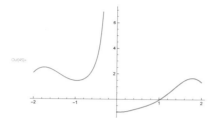

—— Problem 15.8

Investigate $\lim_{n \to 0} \sin(x)/x = 1$.

⟹ SOLUTION.

Clearly the function $\sin(x)/x$ is not defined at 0. We plot the graph to see the behaviour of the function around 0.

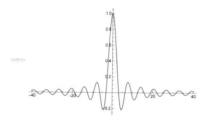

We use `Limit` to have *Mathematica* calculate the limit when x tends to 0.

```
Limit[Sin[x]/x, x -> 0]
1
```

This is in line with what the plot shows us. Next we create a list of numbers which get close to 0 and calculate $\sin(x)/x$ for these numbers. We see that as x tends to 0, $\sin(x)/x$ indeed tends to 1.

```
x = Complement[Range[-1, 1, 0.05], {0.}]

{-1., -0.95, -0.9, -0.85, -0.8, -0.75, -0.7, -0.65, -0.6,
-0.55, -0.5, -0.45, -0.4, -0.35, -0.3, -0.25, -0.2, -0.15,
-0.1, -0.05, 0.05, 0.1, 0.15, 0.2, 0.25, 0.3, 0.35, 0.4, 0.45,
0.5, 0.55, 0.6, 0.65, 0.7, 0.75, 0.8, 0.85, 0.9, 0.95, 1.}
```

```
Sin[x]/x
```

{0.841471, 0.856227, 0.870363, 0.883859, 0.896695, 0.908852,
0.920311, 0.931056, 0.941071, 0.95034, 0.958851, 0.96659,
0.973546, 0.979708, 0.985067, 0.989616, 0.993347, 0.996254,
0.998334, 0.999583, 0.999583, 0.998334, 0.996254, 0.993347,
0.989616, 0.985067, 0.979708, 0.973546, 0.96659, 0.958851,
0.95034, 0.941071, 0.931056, 0.920311, 0.908852, 0.896695,
0.883859, 0.870363, 0.856227, 0.841471}

```
ListPlot[%]
```

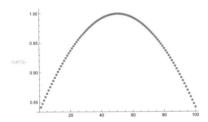

Next we look at derivatives and integrals. The following table give an
overview of how *Mathematica* handles them.

Example	Commands to use
$\frac{\partial f}{\partial x}$	D[f,x]
$\frac{\partial^2 f}{\partial x \partial y}$	D[f,x,y]
$\int f(x)dx$	Integrate[f,x]
$\int_a^b f(x)dx$	Integrate[f,{x,a,b}] or NIntegrate
$\int_c^d \int_a^b f(x,y)dxdy$	Integrate[f[x,y],{y,c,d},{x,a,b}]

―――― Problem 15.9

Evaluate the following:

$$\frac{\partial f}{\partial x}, \text{ when } f = \sin(x)/x,$$

$$\frac{\partial^2 f}{\partial x^2}, \text{ when } f = \sin(x)/x,$$

$$\frac{\partial^3 f}{\partial x^2 \partial y}, \text{ when } f = e^{xy},$$

$$\int \left(\cos(x)/x - \sin(x)/x^2 \right) dx$$

$$\int_{-1}^{1} \int_{-1}^{1} \cos(x^2 + y^2 + xy) dx dy.$$

⟹ SOLUTION.

All we need to do is to translate these into *Mathematica* according to the table above:

```
D[Sin[x]/x, x]
Cos[x]/x - Sin[x]/x^2

D[Sin[x]/x, x, x]
-((2 Cos[x])/x^2) + (2 Sin[x])/x^3 - Sin[x]/x

D[E^(x y), x, x, y]
2 E^(x y) y + E^(x y) x y^2

Integrate[Cos[x]/x - Sin[x]/x^2, x]
Sin[x]/x
```

To evaluate $\int_{-1}^{1} \int_{-1}^{1} \cos(x^2 + y^2 + xy) dx dy$ we can also use Integrate, so *Mathematica* would come up with the precise answer (and it will come up with the precise answer). However, it takes some time:

```
Timing[Integrate[Cos[x^2 + y^2 + x y],
{y, -1, 1}, {x, -1, 1}]][[1]]
21.5496
```

However, if we use NIntegrate, that is, we ask *Mathematica* to approach this calculation numerically, we get the answer in no time.

```
NIntegrate[Cos[x^2 + y^2 + x y], {y, -1, 1}, {x, -1, 1}]
2.81372
```

—— Problem 15.10

Consider the function $\cos(x^2)$. Calculate the area below this function and between 0 and where the function first hits the x-axis.

⟹ SOLUTION.

Here is the graph and the area where we would like to calculate it.

```
Plot[Cos[x^2], {x, 0, Pi}, Filling -> Axis]
```

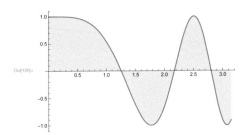

We first find where the function intersects with the x-axis. In order to do so, we solve the equation $\cos(x^2) = 0$, and ask *Mathematica* to find the solution around $x = 1.3$. This can be done using the function FindRoot.

```
FindRoot[Cos[x^2] == 0, {x, 1.3}]
{x -> 1.25331}
```

Once this is done, we find the definite integral to get the desired area.

```
NIntegrate[Cos[x^2], {x, 0, 1.25331}]
0.977451
```

—— Problem 15.11

Consider $f(x, y) = \sin(x + y)\cos(x^2 - y) - \sin(y)$ and generate the graphs of $\frac{\partial^2 f}{\partial x \partial y}$ over the rectangle $-\pi \le x \le \pi$ and $-\pi \le y \le \pi$. Find the maximum of the function $\frac{\partial^2 f}{\partial x \partial y}$ in this area.

⟹ SOLUTION.

We define the function $f(x, y)$ and calculate its second derivative with respect to x and y:

```
f[x_, y_] := Sin[x + y] Cos[x^2 - y] - Sin[y]
```

```
s = D[f[x, y], x, y]
Cos[x + y] Sin[x^2 - y] - 2 x Cos[x + y] Sin[x^2 - y] -
Cos[x^2 - y] Sin[x + y] + 2 x Cos[x^2 - y] Sin[x + y]
```

All we need to do is to plug this equation into `Plot3D` and plot the graph over the rectangle given in the statement of the problem.

```
Plot3D[s, {x, -Pi, Pi}, {y, -Pi, Pi}]
```

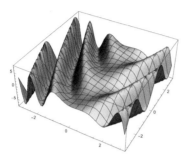

In order to find the maximum of this function, two commands are available, `Maximize`, and its numerical version `NMaximize` which will be much faster and will give you a numerical approximation of the result. There are also two commands for finding the minimum of a function, namely `Minimize` and `NMinimize`.

```
NMaximize[{s, -Pi <= x <= Pi, -Pi <= y <= Pi}, {x, y}]
{7.28319, {x -> -3.14159, y -> 2.57861}}

NMinimize[{s, -Pi <= x <= Pi, -Pi <= y <= Pi}, {x, y}]
{-5.28319, {x -> 3.14159, y -> 2.57861}}
```

These functions give us 7.283 and −5.283 as the maximum and the minimum of this function in the area between $-\pi$ and π. To make sure these values are in line with our expectations, we also plot two planes on these heights.

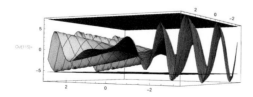

The maximum we obtained using `NMaximize` seems to be ok however the graph of the function clearly goes below the minimum calculated by *Mathe-*

matica. A reason could be that *Mathematica* sometimes only calculates local minima and maxima of functions. Let us concentrate on the area where the graph goes below the plane.

```
Plot3D[{s, -5.283185307179028'}, {x, -Pi, -2}, {y,  Pi, Pi}]
```

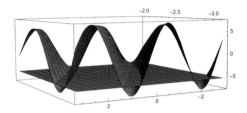

We try to find the minimum in this more restricted area.

```
NMinimize[{s, -Pi <= x <= -2, -Pi <= y <= Pi}, {x, y}]
{-7.28319, {x -> -3.14159, y -> -2.13378}}
```

```
Plot3D[{s, 7.283185307179586', -7.293185258608728'}, {x, -Pi,
    Pi}, {y, -Pi, Pi}, PlotPoints -> 30]
```

The new minimum given is −7.281. We plot the graph with this new height, to make sure the graph is between the planes.

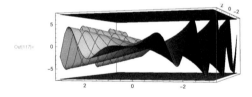

───── Problem 15.12

Consider the following function.

$$f(x) := \sum_{k=1}^{100} \left(\frac{\sin\left(2\pi k^2 x\right)}{4\pi^2 k^5} + \frac{x^2}{2k} \right)$$

This function was suggested by Sungkon Chang as an example of a function that looks quite "innocent" but whose derivatives behave quite wildly. Plot $f(x)$, $f'(x)$ and $f''(x)$ and observe this behaviour.

$\Longrightarrow$ SOLUTION.

We use Palettes, Basic Math Assistance, to enter the function $f(x)$.

$$f(\text{x}\_):=\sum_{k=1}^{100}\left(\frac{\sin(2\pi k^2 x)}{4\pi^2 k^5}+\frac{x^2}{2k}\right)$$

Next we plot the graphs of the function and its derivatives. Note the way we use the Map function (/@).

```
Plot[Evaluate[D[f[x], {x, #}]], {x, 0, 2}] & /@ {0, 1, 2, 3}
```

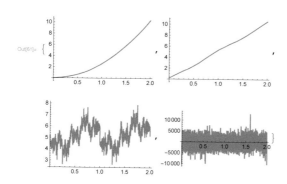

Out[61]=

Problem 15.13

Consider the following functions of two variables $\mathbf{x}(u,v) = \sin(v)\cos(u)$, $\mathbf{y}(u,v) = \sin(v)\sin(u)$ and $\mathbf{z}(u,v) = \cos(v)$. Generate the surface $(\mathbf{x},\mathbf{y},\mathbf{z})$ when $0 \le u \le 3\pi/2$ and $0 \le v \le \pi$. Now consider $\mathbf{x_1}(u,v) = \frac{-3}{8}\cos(v)\sin(\frac{4u}{3})$, $\mathbf{y_1}(u,v) = \frac{3}{8}\cos(\frac{4u}{3})\cos(v)$ and $\mathbf{z_1}(u,v) = \frac{\sin(v)}{2}$. Generate the surface

$$\left(\frac{d\mathbf{x_1}}{dvdu}, \frac{d\mathbf{y_1}}{dudv}, \frac{d\mathbf{z_1}}{dv}\right)$$

when $0 \le u \le 3\pi/2$ and $0 \le v \le \pi$. Finally, superimpose these two images.

$\Longrightarrow$ SOLUTION.

```
x[u_, v_] := Sin[v] Cos[u]
y[u_, v_] := Sin[v] Sin[u]
z[u_, v_] := Cos[v]

ParametricPlot3D[{x[u, v], y[u, v], z[u, v]}, {u, 0, 3 Pi/2},
 {v, 0, Pi}]
```

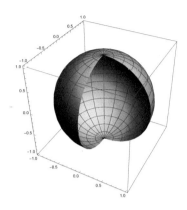

```
x1[u_, v_] := -3/8 Cos[v] Sin[4 u/3]
y1[u_, v_] := 3/8 Cos[4 u/3] Cos[v]
z1[u_, v_] := Sin[v]/2

x2 = D[x1[u, v], u, v]
1/2 Cos[(4 u)/3] Sin[v]

y2 = D[y1[u, v], v, u]
1/2 Sin[(4 u)/3] Sin[v]

z2 = D[z1[u, v], v]
Cos[v]/2

ParametricPlot3D[{x2, y2, z2}, {u, 0, 3 Pi/2}, {v, 0, Pi}]
```

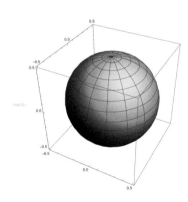

```
Show[Out[334], Out[335]]
```

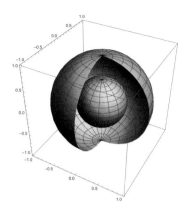

Problem 15.14

Consider two surfaces $q(x,y) = \cos(x^2 + y^2)\exp^{-x^2}$ and $w(x,y) = 3 - x^2 - y^2$. Plot these functions and show the result from the side and bottom view. Find the volume of the region between the graphs on the domain $[-1,1] \times [-1,1]$.

$\Longrightarrow$ SOLUTION.

```
q[x_, y_] := Cos[x^2 + y^2] E^(-x^2)
w[x_, y_] := 3 - x^2 - y^2
Plot3D[{q[x, y], w[x, y]}, {x, -3, 3}, {y, -3, 3}]
```

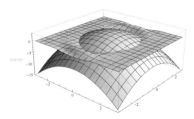

```
NIntegrate[w[x, y] - q[x, y], {x, -1, 1}, {y, -1, 1}]
7.02707
```

Exercise 15.3

Consider the system of equations

$$\begin{cases} x - 2y = -1 \\ (-1+r)x + 3y = 3 \end{cases}$$

First find the unique solution for this system when $r =$. Next explore if r changes, how the solution behaves, and how far the new solution would be from initial solution.

16
Worked out projects

In this chapter we use *Mathematica* to explore mathematics. We look at some interesting topics which mostly appear in discrete mathematics. The problems given here are suitable for student projects.

In this chapter we collect some projects which are suitable as student projects or for self study. The projects are related to some aspects of discrete mathematics and a first year undergraduate training in science will provide enough background to understand and work on them. The codes for the projects are provided, however the reader is encouraged to write their own codes, or to improve (correct) and expand the suggested programs given here.

16.1 Changing of graphs and symbolic dynamics

In this project we study two ways to alter a graph to obtain new graphs. This is related to the subject of symbolic dynamics. A directed graph consists of vertices and edges connecting the vertices to each other. One can think of vertices as cities and edges as direct flights between them. In a more formal language, a *directed graph* $E = (E^0, E^1, r, s)$ consists of two countable sets E^0, E^1 and maps $r, s : E^1 \to E^0$. The elements of E^0 are called *vertices* and the elements of E^1 *edges*. The maps r, s assign the source and target to the edges. If $s^{-1}(v)$ is a finite set for every $v \in E^0$, then the graph is called *row-finite*. In this project we will only consider row-finite graphs. In this setting, if the number of vertices, i.e., $|E^0|$, is finite, then the number of edges, i.e., $|E^1|$, is finite as well and we call E a *finite* graph.

© Springer International Publishing Switzerland 2015
R. Hazrat, *Mathematica®: A Problem-Centered Approach*, Springer Undergraduate Mathematics Series, DOI 10.1007/978-3-319-27585-7_16

For a graph $E = (E^0, E^1, r, s)$, a vertex v for which $s^{-1}(v)$ is empty is called a *sink*, while a vertex w for which $r^{-1}(w)$ is empty is called a *source*. An edge with the same source and range is called a *loop*. A path μ in a graph E is a sequence of edges $\mu = \mu_1 \ldots \mu_k$, such that $r(\mu_i) - s(\mu_{i+1})$, $1 \le i \le k - 1$. In this case, $s(\mu) := s(\mu_1)$ is the *source* of μ, $r(\mu) := r(\mu_k)$ is the *range* of μ, and k is the *length* of μ which is denoted by $|\mu|$. We consider a vertex $v \in E^0$ as a *trivial* path of length zero with $s(v) = r(v) = v$.

Here is an example of a directed graph:

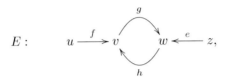

where $E^0 = \{u, v, w, z\}$, $E^1 = \{f, g, h, e\}$ and $s(f) = u$, $r(g) = w$. Here fg and $ghghgh$ are two paths in the graph of length 2 and 6 respectively.

For a graph E, let $n_{v,w}$ be the number of edges with the source v and range w. Then the *adjacency matrix* of the graph E is $A_E = (n_{v,w})$. Usually one orders the vertices and then writes A_E based on this ordering. Two different ordering of vertices give different adjacency matrices. However, if A_E and A'_E are two adjacency matrices of E, then there is a permutation matrix P such that $A'_E = P A_E P^{-1}$.

By giving an adjacency matrix, *Mathematica* can draw the graph associated to the matrix as follows:

```
AdjacencyGraph[( 1 1
                 1 1 ), DirectedEdges → True]
```

Here is a code to draw graphs with edges marked in the graph.

One of the central objects in the theory of symbolic dynamics is a *shift of finite type* (i.e., a topological Markov chain). Every finite directed graph E with no sinks and sources gives rise to a shift of finite type X_E by considering the set of bi-infinite paths and the natural shift of the paths to the left. This is called an *edge shift*. Conversely any shift of finite type is conjugate to an edge shift (for a comprehensive introduction to symbolic dynamics, see the book of Lind

```
el = {1 -> 2, 1 -> 3, 2 -> 3, 2 → 1, 3 → 1, 3 → 2};

AdjacencyGraph[(0 1 1
                1 0 1
                1 1 0), DirectedEdges → True,

  EdgeLabels → Table[el〚i〛 → "e"ᵢ, {i, Length[el]}]]
```

and Marcus[1]). Several invariants have been proposed in order to classify shifts of finite type, among them Williams' (strong) shift equivalence. This is on one hand related to relations between matrices and on the other hand, changes one makes to graphs representing the matrices. They were introduced in an attempt to provide a computable machinery for determining the conjugacy between two shifts of finite type.

Two square nonnegative integer matrices A and B are called *elementary shift equivalent*, denoted by $A \sim_{ES} B$, if there are nonnegative matrices R and S such that $A = RS$ and $B = SR$.

Example 16.1

Let $A = (2)$, and $B = \begin{pmatrix} 1 & 1 \\ 1 & 1 \end{pmatrix}$ be two matrices. Then $A \sim_{ES} B$ as

$$(2) = (1 \quad 1) \begin{pmatrix} 1 \\ 1 \end{pmatrix},$$

$$\begin{pmatrix} 1 & 1 \\ 1 & 1 \end{pmatrix} = \begin{pmatrix} 1 \\ 1 \end{pmatrix} (1 \quad 1).$$

The equivalence relation $\sim_S$ on square nonnegative integer matrices generated by elementary shift equivalence is called *strong shift equivalence*. That is, $A \sim_S B$ if

$$A = A_0 \sim_{ES} A_1 \sim_{ES} A_2 \sim_{ES} \cdots \sim_{ES} A_n = B.$$

---

[1] D. Lind and B. Marcus. An introduction to symbolic dynamics and coding, Cambridge University Press, 1995.

Example 16.2

Let $A = \begin{pmatrix} 1 & 2 \\ 1 & 0 \end{pmatrix}$ and $A^T = \begin{pmatrix} 1 & 1 \\ 2 & 0 \end{pmatrix}$. We show that A is strongly shift equivalent to A^T. We have

$$A = \begin{pmatrix} 1 & 1 & 0 \\ 0 & 0 & 1 \end{pmatrix} \begin{pmatrix} 1 & 1 \\ 0 & 1 \\ 1 & 0 \end{pmatrix},$$

$$\begin{pmatrix} 1 & 1 \\ 0 & 1 \\ 1 & 0 \end{pmatrix} \begin{pmatrix} 1 & 1 & 0 \\ 0 & 0 & 1 \end{pmatrix} = \begin{pmatrix} 1 & 1 & 1 \\ 0 & 0 & 1 \\ 1 & 1 & 0 \end{pmatrix} = E_1.$$

Furthermore, we have

$$E_1 = \begin{pmatrix} 1 & 1 & 1 \\ 0 & 0 & 1 \\ 1 & 1 & 0 \end{pmatrix} = \begin{pmatrix} 0 & 1 & 1 \\ 1 & 0 & 0 \\ 0 & 0 & 1 \end{pmatrix} \begin{pmatrix} 0 & 0 & 1 \\ 0 & 0 & 1 \\ 1 & 1 & 0 \end{pmatrix},$$

$$\begin{pmatrix} 0 & 0 & 1 \\ 0 & 0 & 1 \\ 1 & 1 & 0 \end{pmatrix} \begin{pmatrix} 0 & 1 & 1 \\ 1 & 0 & 0 \\ 0 & 0 & 1 \end{pmatrix} = \begin{pmatrix} 0 & 0 & 1 \\ 0 & 0 & 1 \\ 1 & 1 & 1 \end{pmatrix} = E_2.$$

Finally,

$$E_2 = \begin{pmatrix} 0 & 0 & 1 \\ 0 & 0 & 1 \\ 1 & 1 & 1 \end{pmatrix} = \begin{pmatrix} 1 & 0 \\ 1 & 0 \\ 1 & 1 \end{pmatrix} \begin{pmatrix} 0 & 0 & 1 \\ 1 & 1 & 0 \end{pmatrix},$$

$$\begin{pmatrix} 0 & 0 & 1 \\ 1 & 1 & 0 \end{pmatrix} \begin{pmatrix} 1 & 0 \\ 1 & 0 \\ 1 & 1 \end{pmatrix} = \begin{pmatrix} 1 & 1 \\ 2 & 0 \end{pmatrix} = A^T.$$

This shows that

$$A \sim_{ES} E_1 \sim_{ES} E_2 \sim_{ES} A^T.$$

Thus $A \sim_S A^T$.

Besides elementary and strong shift equivalence, there is a weaker notion, called shift equivalence, defined as follows. The nonnegative integer square matrices A and B are called *shift equivalent* if there are nonnegative matrices R and S such that $A^l = RS$ and $B^l = SR$, for some $l \in \mathbb{N}$, and $AR = RB$ and $SA = BS$. Clearly the strong shift equivalence implies shift equivalence, but the converse, an open question for almost 20 years, does not hold.

We describe the concept of out-splitting and in-splitting of a graph. The aim of this project is to write codes to perform out-splitting and in-splitting of graphs.

Definition 16.3

Let $E = (E^0, E^1, r, s)$ be a finite graph. For each $v \in E^0$ which is not a sink, partition $s^{-1}(v)$ into disjoint nonempty subsets $\mathcal{E}_v^1, \ldots, \mathcal{E}_v^{m(v)}$, where $m(v) \geq 1$. (If v is a sink then put $m(v) = 0$.) Let $\mathcal{P}$ denote the resulting partition of E^1. We form the *out-split graph* $E_s(\mathcal{P})$ *from* E *using* $\mathcal{P}$ as follows: Let

$$
\begin{aligned}
E_s(\mathcal{P})^0 &= \{v^i \mid v \in E^0, 1 \leq i \leq m(v)\} \cup \{v \mid m(v) = 0\}, \\
E_s(\mathcal{P})^1 &= \{e^j \mid e \in E^1, 1 \leq j \leq m(r(e))\} \cup \{e \mid m(r(e)) = 0\},
\end{aligned}
$$

and define $r_{E_s(\mathcal{P})}, s_{E_s(\mathcal{P})} : E_s(\mathcal{P})^1 \to E_s(\mathcal{P})^0$ for $e \in \mathcal{E}_{s(e)}^i$ by

$$
\begin{aligned}
s_{E_s(\mathcal{P})}(e^j) &= s(e)^i &&\text{and} && s_{E_s(\mathcal{P})}(e) = s(e)^i, \\
r_{E_s(\mathcal{P})}(e^j) &= r(e)^j &&\text{and} && r_{E_s(\mathcal{P})}(e) = r(e).
\end{aligned}
$$

Example 16.4

Consider the graph

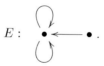

$$E :$$

Let $\mathcal{P}$ be the partition of the edges of E containing only one edge in each partition. Then the out-split graph of E using $\mathcal{P}$ is

$$E_s(\mathcal{P}) :$$

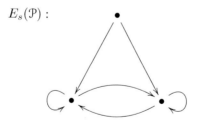

Here is another example of out-splitting. We start with the graph on page 260 with two vertices and 4 edges, partition the edges such that each set contains only one edge, and apply the out-splitting process. We then use the adjacency matrix of the result to ask *Mathematica* to plot the new graph.

The in-splitting of a graph is of the same nature as the out-splitting, as the following definition shows.

`AdjacencyGraph[`$\left(\begin{smallmatrix} 1 & 1 \\ 1 & 1 \end{smallmatrix}\right)$`, DirectedEdges → True]`

`AdjacencyGraph[`$\left(\begin{smallmatrix} 1 & 1 & 0 & 0 \\ 0 & 0 & 1 & 1 \\ 0 & 0 & 1 & 1 \\ 1 & 1 & 0 & 0 \end{smallmatrix}\right)$`, DirectedEdges → True]`

Definition 16.5

Let $E = (E^0, E^1, r, s)$ be a finite graph. For each $v \in E^0$ which is not a source, partition the set $r^{-1}(v)$ into disjoint nonempty subsets $\mathcal{E}_1^v, \ldots, \mathcal{E}_{m(v)}^v$, $m(v) \geq 1$. (If v is a source then put $m(v) = 0$.) Let $\mathcal{P}$ denote the resulting partition of E^1. We form the *in-split graph* $E_r(\mathcal{P})$ from E using $\mathcal{P}$ as follows: Let

$$E_r(\mathcal{P})^0 = \{v_i \mid v \in E^0, 1 \leq i \leq m(v)\} \cup \{v \mid m(v) = 0\},$$
$$E_r(\mathcal{P})^1 = \{e_j \mid e \in E^1, 1 \leq j \leq m(s(e))\} \cup \{e \mid m(s(e)) = 0\},$$

and define $r_{E_r(\mathcal{P})}, s_{E_r(\mathcal{P})} : E_r(\mathcal{P})^1 \to E_r(\mathcal{P})^0$ for $e \in \mathcal{E}_i^{r(e)}$ by

$$s_{E_r(\mathcal{P})}(e_j) = s(e)_j \quad \text{and} \quad s_{E_r(\mathcal{P})}(e) = s(e),$$
$$r_{E_r(\mathcal{P})}(e_j) = r(e)_i \quad \text{and} \quad r_{E_r(\mathcal{P})}(e) = r(e)_i.$$

Theorem 16.6 (WILLIAMS)

Let A and B be two square nonnegative integer matrices and let E and F be their associated graphs. Then A is strongly shift equivalent to B if and only if E can be obtained from F by a sequence of in/out-splittings and their converses.

We write a code to perform the out-splitting of any graph, using the maximum partitioning, i.e., each edge is in one partition set. We leave it to the reader to decipher the code, improve it and from there write a code for the in-splitting case.

```
outSplit[m_] := Module[{i, j, k, lm = Length[m], bb1, bb2},
  rowadd[p_, q_] := Plus @@ Flatten[p[[Range[q]]]];
  d = Plus @@ Flatten[m];
  a = Table[0, {d}, {d}];
  Do[
    bb1 = Range[1 + rowadd[m, i - 1] + Sum[m[[i, k]],
    {k, 1, j - 1}], rowadd[m, i - 1] + Sum[m[[i, k]], {k, 1, j}]];
    bb2 = Range[1 + rowadd[m, j - 1], rowadd[m, j]];
    a[[bb1, bb2]] = 1,
    {j, 1, lm}, {i, 1, lm}];
  a]
```

$$b = \begin{pmatrix} 1 & 1 \\ 1 & 1 \end{pmatrix};$$

outSplit[b] // MatrixForm

```
rixForm=
( 1 1 0 0 )
( 0 0 1 1 )
( 1 1 0 0 )
( 0 0 1 1 )
```

AdjacencyGraph[outSplit[b]]

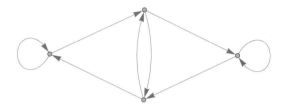

As is clear from the code, we identify the graph with its adjacency matrix. The code outSplit returns a matrix which is the out-splitting of the original matrix. We can then visualise this with the command AdjacencyGraph.

Already for 2×2 matrices (i.e., graphs with two matrices) the out-splitting procedure generates interesting graphs.

$$m = \begin{pmatrix} 1 & 2 \\ 8 & 0 \end{pmatrix};$$

`AdjacencyGraph[outSplit[m]]`

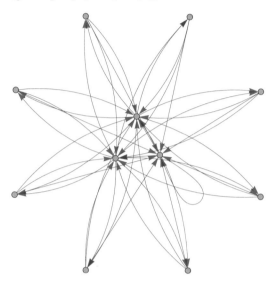

`AdjacencyGraph[Nest[outSplit, m, 5]]`

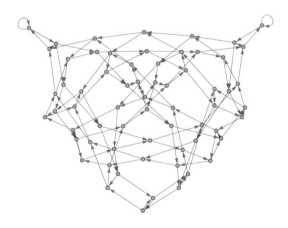

16.2 Partitioned binary relations

Let X and Y be sets. A *binary relation* between A and B is a subset of $A \times B$. Informally, this means we are relating some elements of A and B to each other. Starting from a binary relation, one defines a binary operation, that is we assign a new element to each pair of elements of A and B. For example, consider all the pairs (m, n), where $m, n \in \mathbb{Z}$. This defines a binary relation between $\mathbb{Z}$ and $\mathbb{Z}$. In this case the binary relation is in fact the set $\mathbb{Z} \times \mathbb{Z}$. The addition as a binary operation is then defined for each pair (m, n) by assigning $m + n$ to the pair.

Martin and Mazorchuk[2] define a very general binary relations between sets and then define an operation to compose the elements. Following their paper, we first define the partitioned binary relation and then write a code to compose the relations.

Let X and Y be finite sets. A *partitioned binary relation* on (X, Y) is a binary relation α on the *disjoint* union of X and Y. The sets X and Y are called the *domain* and the *codomain* of α and denoted by $\mathrm{Dom}(\alpha)$ and $\mathrm{Codom}(\alpha)$, respectively.

A partitioned binary relation α on (X, Y) will be depicted as a directed graph drawn within a rectangular frame, with elements of X and Y represented by vertexes positioned on the right- and left-hand sides of the frame, respectively. The fact that α contains an edge $(a, b) \in (X \coprod Y)^2$ will be written $(a, b) \in \alpha$ and visualised by an arrow from a to b on the graph. We will call a and b *elements* while (a, b) will be called an *edge*. An example of a partitioned binary relation from $X = \{x_1, x_2, x_3, x_4\}$ to $Y = \{y_1, y_2, y_3, y_4, y_5, y_6, y_7, y_8, y_9\}$ is shown in Figure 16.1.

Next we define the composition of partitioned binary relations. Namely, given a partitioned binary relation α on (X, Y) and a partitioned binary relation β on (Y, Z), we define their composition $\beta \circ \alpha$, which will be a partitioned binary relation on (X, Z).

Let $\aleph = (\alpha_1, \alpha_2)$ be a *composable* partitioned binary relation in the above sense, that is $\mathrm{Codom}(\alpha_1) = \mathrm{Dom}(\alpha_2)$. Set $X_i := \mathrm{Dom}(\alpha_i)$ for $i = 1, 2$, $X_3 := \mathrm{Codom}(\alpha_2)$, and $X_{\coprod} := \coprod_{i=1}^{3} X_i$. A sequence $\xi = (a_1, b_1), (a_2, b_2), \dots, (a_m, b_m)$ of edges taken from the partitioned binary relations in $\aleph$ is called $\aleph$-*connected* provided that

1. no two successive edges in ξ are in the same partitioned binary relation;

2. for every $i = 1, 2, \dots, m - 1$ we have $b_i = a_{i+1}$ (as elements of $X_{\coprod}$).

---

[2] P. Martin, V. Mazorchuk, Partitioned binary relations, arXiv:1102.0862.

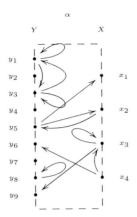

Figure 16.1   A partitioned binary relation on (X, Y)

We will also say that the $\aleph$-connected sequence ξ *connects* a_1 to b_m. Note that at each step i the element b_i defines the partitioned binary relation α_j containing (a_{i+1}, b_{i+1}) uniquely due to condition (2). Note also that in the case $k = 1$, we necessarily have $m = 1$.

Let α be a partitioned binary relation on (X, Y) and β be a partitioned binary relation on (Y, Z). We define the *composition* $\beta \circ \alpha$ as the partitioned binary relation on (X, Z) such that for every $a, b \in X \coprod Z$ the partitioned binary relation $\beta \circ \alpha$ contains (a, b) if and only if there exists an (α, β)-connected sequence connecting a to b. An example of the composition of two partitioned binary relations is shown in Figure 16.2.

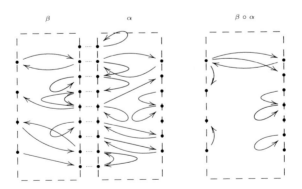

Figure 16.2   Composition of partitioned binary relations

In their paper Martin and Mazorchuk study this binary relation and obtain known relations as special cases of this general definition. As one can see from

Figure 16.2, the composition is quite involved. We would like to write a program to produce this composition for us.

We run the code for the example given in Figure 16.2. We label the elements as follows: $X = \{1, 2, \ldots, 7\}$, $Y = \{8, 9, \ldots, 16\}$ and $Z = \{17, 18, \ldots, 20\}$. Now from Figure 16.2, it is clear that

$$
\beta = \begin{pmatrix} 9 & 17 \\ 17 & 9 \\ 10 & 10 \\ 11 & 12 \\ 12 & 11 \\ 13 & 13 \\ 12 & 18 \\ 15 & 19 \\ 19 & 15 \\ 20 & 16 \end{pmatrix}, \quad
\alpha = \begin{pmatrix} 1 & 9 \\ 9 & 1 \\ 8 & 8 \\ 9 & 10 \\ 10 & 11 \\ 12 & 12 \\ 12 & 2 \\ 3 & 4 \\ 4 & 4 \\ 13 & 6 \\ 6 & 13 \\ 14 & 7 \\ 7 & 14 \\ 15 & 16 \\ 16 & 15 \end{pmatrix}.
$$

We represent α and β with lists in *Mathematica* as follows.

$$\alpha = \{\{1, 9\}, \{9, 1\}, \{8, 8\}, \{9, 10\}, \{10, 11\}, \{12, 12\},$$
$$\{12, 2\}, \{3, 4\}, \{4, 4\}, \{13, 6\}, \{6, 13\}, \{14, 7\}, \{7, 14\}, \{15, 16\}, \{16, 15\}\},$$

and

$$\beta = \{\{9, 17\}, \{17, 9\}, \{10, 10\}, \{11, 12\}, \{12, 11\}, \{13, 13\},$$
$$\{12, 18\}, \{15, 19\}, \{19, 15\}, \{20, 16\}\}.$$

```
X = Range[1, 7]
{1, 2, 3, 4, 5, 6, 7}

Y = Range[8, 16]
{8, 9, 10, 11, 12, 13, 14, 15, 16}

Z = Range[17, 20]
{17, 18, 19, 20}

A = {{1, 9}, {9, 1}, {8, 8}, {9, 10}, {10, 11},
{12, 12}, {12, 2}, {3, 4}, {4, 4}, {13, 6}, {6, 13},
{14, 7}, {7, 14}, {15, 16}, {16, 15}};

B = {{9, 17}, {17, 9}, {10, 10}, {11, 12}, {12, 11},
{13, 13}, {12, 18}, {15, 19}, {19, 15}, {20, 16}};
```

Here is the whole code, which consists of two recursive functions. The main recursive function p takes care of all the combinations of compositions of sequences of pairs appearing in α and β.

```
p[{a_, b_?(MemberQ[X, #] &), c_: 1}] :- {a, b, c}
p[{a_, b_?(MemberQ[Z, #] &), c_: 2}] := {a, b, c}
p[{a_, b_, c_?(# > 20 &)}] := "loop"

p[x : {{_, _, _} ..}] := DeleteCases[Level[p /@ x, {-2}], {}]

p[{a_?(MemberQ[X, #] &), b_?(MemberQ[Y, #] &)}] :=
    p[ Cases[B, {b, x_} -> {b, x, 2}]]

p[{a_?(MemberQ[Z, #] &), b_?(MemberQ[Y, #] &)}] :=
    p[Cases[A, {b, x_} -> {b, x, 1}]]

p[{a_?(MemberQ[Y, #] &), b_?(MemberQ[Y, #] &), aa_?OddQ}] :=
    p[Cases[B, {b, x_} -> {b, x, aa + 1}]]

p[{a_?(MemberQ[Y, #] &), b_?(MemberQ[Y, #] &), bb_?EvenQ}] :=
    p[Cases[A, {b, x_} -> {b, x, bb + 1}]]

q[{a_, b_}] := p[{a, b}] /. {x_Integer, y_, z_} :> {a, y}

fp[s_] := Union[Cases[
Level[q /@s, {-2}], {x_?(!MemberQ[Y, #] &), y_Integer}->{x, y}]]

fp[x_, y_] := Flatten[fp /@ {x, y}, 1]
```

Now plugging these into the program we get:

```
fp[A, B]

{{1, 17}, {3, 4}, {4, 4}, {6, 6}, {17, 1}, {17, 2},
 {17, 18}, {20, 19}}
```

Thus we have

$$\beta \circ \alpha = \begin{pmatrix} 1 & 17 \\ 3 & 4 \\ 4 & 4 \\ 6 & 6 \\ 17 & 1 \\ 17 & 2 \\ 17 & 18 \\ 20 & 19 \end{pmatrix},$$

which coincides with what is shown in Figure 16.2.

Here is another example.

```
X = Range[1, 4]
{1, 2, 3, 4}

Y = Range[5, 8]
{5, 6, 7, 8}

Z = Range[9, 12]
{9, 10, 11, 12}

A = {{6, 1}, {6, 7}, {2, 7}, {8, 7}, {8, 4}}

B = {{9, 6}, {7, 6}, {7, 8}, {8, 12}}

fp[A, B]
{{2, 1}, {2, 4}, {9, 1}, {9, 4}}
```

Let us analyse the code. We start we a pair (a, b) in α. Since the pair comes from the disjoint union of X and Y, a and b could be from any of these sets. If we want to compose this with an element of β, say (d, e), then, by the regulation (see 1), d has to be in Y. This means if it happens that $b \in X$, then there is no way to compose this with any other pair, and thus composition stops here. (To clarify, if we were working with say (X, X), then in the disjoint union, the first X is distinguished from the second X and thus we are allowed to talk about different X's.) This is why we define

```
p[{a_, b_?(MemberQ[X, #] &), c_: 1}] := {a, b, c}
```

The third argument c is designed to keep track whether the pair is in α or β. We discuss the role of c later. Similarly if in the pair (d, e), e happens to be in Z, then one cannot compose any further, and thus

```
p[{a_, b_?(MemberQ[Z, #] &), c_: 2}] := {a, b, c}
```

Now if (a, b) is in α and $b \in Y$, then we collect all the pairs (b, d) in β:

```
p[{a_?(MemberQ[X, #] &), b_?(MemberQ[Y, #] &)}] :=
    p[ Cases[B, {b, x_} -> {b, x, 2}]]
```

Since it maybe the case that the result is a pair (d, e) where $d, e \in Y$, and we need to determine if this pair is in α or is in β, we assign to c an even number (here 2) if the pair is in β and odd number if the pair is in α.

Similarly if (a, b) is in β and $b \in Y$, we collect all (b, d) where $d \in X$. These are done with the following functions.

```
p[{a_?(MemberQ[Z, #] &), b_?(MemberQ[Y, #] &)}] :=
    p[Cases[A, {b, x_} -> {b, x, 1}]]
```

Again note that here the third component c is assigned 1, an odd number. Next we take care of the cases where in the pair (a, b), both $a, b \in Y$.

If the pair is in α, and this would be clear if the third component here aa is odd, then we compose them with elements from β.

```
p[{a_?(MemberQ[Y, #] &), b_?(MemberQ[Y, #] &), aa_?OddQ}] :=
  p[Cases[B, {b, x_} -> {b, x, aa + 1}]]
```

But if c is even, then (a, b) is already in β and thus we compose them with the elements of α.

```
p[{a_?(MemberQ[Y, #] &), b_?(MemberQ[Y, #] &), bb_?EvenQ}] :=
  p[Cases[A, {b, x_} -> {b, x, bb + 1}]]
```

Now the rest is recursive, i.e., we repeat the above process. The only case that one has to take into account is if this recursive function ends up in a loop, and this is possible if, say $(y_1, y_2) \in \alpha$ where $y_1, y_2 \in Y$ and $(y_2, y_1) \in \beta$. Our code will keep repeating the composition and each time increases the third component c. Thus we design the code

```
p[{a_, b_, c_?(# > 20 &)}] := "loop"
```

This detects if there has been pairs such that composition has created a loop (and thus increases c). If this goes beyond certain limit, we stop. Here we assign a limit of 20, however if we are working with large sets for X, Y and Z, it might be necessary to consider a larger bound for c. The rest of the code should be easy to decipher and we leave it to the reader.

16.3 Persian recursions

The following amusing algorithm was suggested in a short paper by Anne Burns to create "Persian rug-like pattern" using a recursive procedure.[3] Following Burns' paper, we describe the procedure here and then we will write a *Mathematica* code to generate some of these patterns.

Consider a $2^n + 1 \times 2^n + 1$ celled square. The aim is to colour each cell based on the colour of the corners already coloured. This is achieved by allocating a number between 0 and $m - 1$ to each cell, m being the number of available colours. We begin to colour the outer cells first, colouring them all the same colour in order to form a border. We then apply the following procedure recursively:

1. Use the four cells in the corners and a four variable function to determine a different colour. For example the function could be

$$f(c_1, c_2, c_3, c_4) = (c_1 + c_2 + c_3 + c_4) \mod m.$$

2. Allocate the newly determined colour to the interior row cells and column cells in the centre.

3. Apply this scheme to all of the four new bordered squares generated at each execution.

Once all cells have been coloured the function will terminate. As an example, for $2^3 + 1 \times 2^3 + 1$ cells with $m = 7$ and the boundary cells all 4, by following the above procedure, we get the following matrix.

$$\begin{pmatrix} 4 & 4 & 4 & 4 & 4 & 4 & 4 & 4 & 4 \\ 4 & 5 & 0 & 3 & 2 & 3 & 0 & 5 & 4 \\ 4 & 0 & 0 & 0 & 2 & 0 & 0 & 0 & 4 \\ 4 & 3 & 0 & 6 & 2 & 6 & 0 & 3 & 4 \\ 4 & 2 & 2 & 2 & 2 & 2 & 2 & 2 & 4 \\ 4 & 3 & 0 & 6 & 2 & 6 & 0 & 3 & 4 \\ 4 & 0 & 0 & 0 & 2 & 0 & 0 & 0 & 4 \\ 4 & 5 & 0 & 3 & 2 & 3 & 0 & 5 & 4 \\ 4 & 4 & 4 & 4 & 4 & 4 & 4 & 4 & 4 \end{pmatrix}.$$

We first translate the procedure into *Mathematica*, writing a code to generate a matrix with the colours assigned to the entries. This requires a recursive function, finding the middle row and column and then calling the function again four times for each of the four new matrices after the partition.

---

[3] Anne M. Burns, "Persian" recursion, Mathematics Magazine, Vol. 70, No. 3, 1997, 196-199

To start with, we consider the matrix x and eventually fill all the entries with the specific colours.

```
s=2^3+1
```

```
x = Array[n, {s, s}]
```

First we assign the colour 4 to the boundaries (see Problem 4.3).

```
x[[All, 1]] = x[[1, All]] = x[[-1, All]] = x[[All, -1]] = 4
```

Next we define the function cx[l_, r_, t_, b_], of step one of the procedure, where l, r, t and b are the left, right, top and bottom coordinates of the matrix.

```
cx[l_, r_, t_, b_] :=
    Mod[(x[[t, 1]] + x[[t, r]] + x[[b, 1]] + x[[b, r]]), 7];
```

With mc = (1 + r)/2, mr = (t + b)/2 we find the middle rows and column and the cx gives the new colour. This can be done by

```
x[[mr, 1 + 1 ;; r - 1]] =
    x[[t + 1 ;; b - 1, mc]] = cx[l, r, t, b];
```

$$\begin{pmatrix} 4 & 4 & 4 & 4 & 4 & 4 & 4 & 4 & 4 \\ 4 & \square & \square & \square & 2 & \square & \square & \square & 4 \\ 4 & \square & \square & \square & 2 & \square & \square & \square & 4 \\ 4 & \square & \square & \square & 2 & \square & \square & \square & 4 \\ 4 & 2 & 2 & 2 & 2 & 2 & 2 & 2 & 4 \\ 4 & \square & \square & \square & 2 & \square & \square & \square & 4 \\ 4 & \square & \square & \square & 2 & \square & \square & \square & 4 \\ 4 & \square & \square & \square & 2 & \square & \square & \square & 4 \\ 4 & 4 & 4 & 4 & 4 & 4 & 4 & 4 & 4 \end{pmatrix}$$

The function colorGrid[l_, r_, t_, b_] will assign the new colours to the entries of the matrix.

```
colorGrid[l_, r_, t_, b_] := colorGrid[l, r, t, b] =
    With[{mc = (1 + r)/2, mr = (t + b)/2},
      If[l < r - 1,
      x[[mr, 1 + 1 ;; r - 1]] =
        x[[t + 1 ;; b - 1, mc]] = cx[l, r, t, b];
      colorGrid[l, mc, t, mr];
      colorGrid[mc, r, t, mr];
      colorGrid[l, mc, mr, b];
      colorGrid[mc, r, mr, b]
      ]];
    x[[All, 1]] = x[[1, All]] = x[[-1, All]] = x[[All, -1]] = 4;
    colorGrid[1, s, 1, s]; x
    ]
```

Note that, within `colorGird` we have called the function four times to take care of the four squares created after partitioning the matrix.

Now we are in a position to put all these codes together. We use a `Module` to put all the codes under one roof.

```
perCarpet[s_, m_, z_] := Module[{n, l, r, t, b, colorGrid,
  x = Array[n, {s, s}]},
 cx[l_, r_, t_, b_] :=
 Mod[((x[[t, l]] + x[[t, r]] + x[[b, l]] + x[[b, r]]), m];
 colorGrid[l_, r_, t_, b_] := colorGrid[l, r, t, b] =
  With[{mc = (l + r)/2, mr = (t + b)/2},
   If[l < r - 1,
    x[[mr, l + 1 ;; r - 1]] =
    x[[t + 1 ;; b - 1, mc]] = cx[l, r, t, b];
    colorGrid[l, mc, t, mr];
    colorGrid[mc, r, t, mr];
    colorGrid[l, mc, mr, b];
    colorGrid[mc, r, mr, b]
    ]];
 x[[All, 1]] = x[[1, All]] = x[[-1, All]] = x[[All, -1]] = z;
  colorGrid[1, s, 1, s]; x
 ]
```

We try the function with different values for n, m and the boundary colours.

`l = perCarpet[2^3 + 1, 14, 5] // MatrixForm`

$$\begin{pmatrix} 5 & 5 & 5 & 5 & 5 & 5 & 5 & 5 & 5 \\ 5 & 8 & 7 & 9 & 6 & 9 & 7 & 8 & 5 \\ 5 & 7 & 7 & 7 & 6 & 7 & 7 & 7 & 5 \\ 5 & 9 & 7 & 11 & 6 & 11 & 7 & 9 & 5 \\ 5 & 6 & 6 & 6 & 6 & 6 & 6 & 6 & 5 \\ 5 & 9 & 7 & 11 & 6 & 11 & 7 & 9 & 5 \\ 5 & 7 & 7 & 7 & 6 & 7 & 7 & 7 & 5 \\ 5 & 8 & 7 & 9 & 6 & 9 & 7 & 8 & 5 \\ 5 & 5 & 5 & 5 & 5 & 5 & 5 & 5 & 5 \end{pmatrix}.$$

`l = perCarpet[2^3 + 1, 9, 4] // MatrixForm`

$$\begin{pmatrix} 4 & 4 & 4 & 4 & 4 & 4 & 4 & 4 & 4 \\ 4 & 4 & 1 & 7 & 7 & 7 & 1 & 4 & 4 \\ 4 & 1 & 1 & 1 & 7 & 1 & 1 & 1 & 4 \\ 4 & 7 & 1 & 4 & 7 & 4 & 1 & 7 & 4 \\ 4 & 7 & 7 & 7 & 7 & 7 & 7 & 7 & 4 \\ 4 & 7 & 1 & 4 & 7 & 4 & 1 & 7 & 4 \\ 4 & 1 & 1 & 1 & 7 & 1 & 1 & 1 & 4 \\ 4 & 4 & 1 & 7 & 7 & 7 & 1 & 4 & 4 \\ 4 & 4 & 4 & 4 & 4 & 4 & 4 & 4 & 4 \end{pmatrix}.$$

Now that we have the matrix with all the colours assigned (i.e., all the entries determined), the next step is to replace each cell with a rectangle with the assigned colour. We first get all the colours which appear in the cells of x by Rest[Union[Flatten[x]]]. We define a set with the colours as follows:

```
c = {Red, Blue, Black, White, Gray, Cyan, Magenta, Yellow,
Brown, Orange, Pink, Purple, LightRed, LightBlue, LightYellow,
Green};
```

The function d below finds the coordinates of the cells with a given colour and adds that colour to the entries, so that we can pass all these into the function Rectangle.

```
d[n_] := Outer[List, {c[[n]]}, Position[x, n], 1]
```

Now we are ready to replace the entries with the Rectangle and then call on Graphics to plot them together.

```
v = Union[
    Map[d, Rest[Union[Flatten[x]]]]] /. {n_, {k_, l_}} ->
    {n, Rectangle[{k, l}, {k + 1, l + 1}]};

Graphics[Flatten[v, 2]]
```

For $2^3 + 1$ we get

```
x = perCarpet[2^3 + 1];
```

and the carpet will look like this:

For $2^8 + 1$ we get:

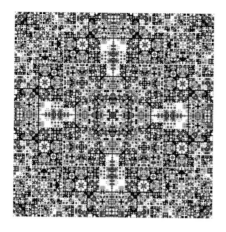

Changing `c[x]` to

```
Mod[Floor[(x[[t, l]] + x[[t, r]] + x[[b, l]] +
   x[[b, r]])/3], 16]}
```

and the boundary colour to 1, we get:

There is an alternative way to draw the graph using `ArrayPlot`.

```
ArrayPlot[Table[Sin[x^2 + y^2], {x, -40, 40}, {y, -40, 40}]]
```

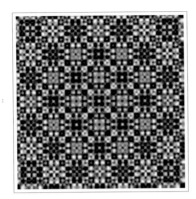

In order to use colours in `ArrayPlot` we need to assign them to the numbers in the entries via a `Rule`, as follows (see Exercise 8.8).

```
Inner[Rule, Range[16], c, List]
```

$\{1 \to \blacksquare, 2 \to \blacksquare, 3 \to \blacksquare, 4 \to \square, 5 \to \blacksquare, 6 \to \square, 7 \to \blacksquare, 8 \to \square,$
$9 \to \blacksquare, 10 \to \blacksquare, 11 \to \blacksquare, 12 \to \blacksquare, 13 \to \square, 14 \to \square, 15 \to \square, 16 \to \blacksquare\}$

Now we are ready to use this alternative (and rather more elegant way) to colour the matrix.

```
ArrayPlot[perCarpet[2^8 + 1, 9, 4], ColorRules -> col]
```

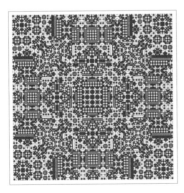

17
Projects

Here is a list of projects readers/students can choose to work on in the lab. The students are encouraged to "investigate" the projects, change the problems and produce their own projects out of them. Projects A are short interesting problems that require rather short codes to achieve the desired outcome. Projects B require rather longer codes to achieve the desired outcome. The total marks will be awarded, split between the correctness of the project (75%) and presentation style (25% marks). The latter incorporates items such as clarity of code and the exposition.

17.1 Projects A

1. Suppose P_i is the i-th prime number. Call $E_i = P_1 \times P_2 \times \cdots \times P_i + 1$ the i-th Euclid number. Define a function `primeProduct` recursively to produce the product of the first n primes. Use this function and produce the list of the first 20 Euclid Numbers. Find which of these numbers are in fact prime. Find out the indices of the prime Euclid numbers (i.e, i where E_i is prime) where $1 \leq i \leq 100$.

2. Let us call a number $a_0 a_1 ... a_{n-1} a_n$ fully seven-divisible if $a_0 a_1 ... a_{n-1} a_n$, $a_0 a_1 ... a_{n-1}, \cdots, a_0 a_1$ and a_0 are all divisible by 7 (e.g. 707 is a fully seven divisible as 707, 70 and 7 are all divisible by 7). Find all the 3, 4 and 5 digits numbers which are fully seven-divisible. Can you conjecture how

© Springer International Publishing Switzerland 2015
R. Hazrat, *Mathematica®: A Problem-Centered Approach*, Springer Undergraduate
Mathematics Series, DOI 10.1007/978-3-319-27585-7_17

many fully seven divisible 10 digit numbers exist?

3. Consider two surfaces $q(x, y) = \sin(x^2 + y^2)\exp^{-x^2+x}$ and $w(x, y) = 4 - x^2 - y^2$. Plot both of these functions on the same graph for $-\pi \le x, y \le \pi$ and show the result from both side and bottom views. Find the volume of the region between the graphs on the domain $[-1, 1] \times [-1, 1]$.

4. Investigate and determine the values of x between $0 \le x \le 2\pi$ such that we have the inequality

$$2\cos(x) \le \sqrt{1 + \sin(2x)} - \sqrt{1 - \sin(2x)} \le \sqrt{2}.$$

5. **a.** Write the function

$$e(n) = 1 + \frac{1}{1!} + \frac{1}{2!} + \frac{1}{3!} + \cdots + \frac{1}{n!}$$

in *Mathematica* and show that $e(\infty) = E$, the exponential constant.

b. Let F_i be the $i - th$ Fibonacci number. Write the function

$$f(n) = (F_1 + x)(F_2 + x^2) \cdots (F_n + x^n).$$

What is the coefficient of x^4 in $f(23)$?

6. Define the functions

$$f(t, a) = 2 + \frac{1}{2}\sin(at)$$

$$g(t, b, c) = \cos(t + \frac{\sin(bt)}{c})$$

$$h(t, b, c) = \sin(t + \frac{\sin(bt)}{c}).$$

Generate the parametric graph

$$x(t) = f(t, 8)g(t, 16, 4)$$

$$y(t) = f(t, 8)h(t, 16, 4)$$

when $0 \le t \le 2\pi$. Also plot the graph

$$(f(t, 6)g(t, 18, 18), f(t, 6)h(t, 18, 18))$$

when $0 \le t \le 2\pi$.

7. Plot the graphs of the functions $2\exp^{-x^2}$ and $\cos(\sin(x) + \cos(x))$ between $[-\pi, \pi]$. Then find out the coordinates x where they intersect.

8. Let M be a positive integer and consider the set

$$S = \{n \in \mathbb{N} \mid M^2 \le n \le (M+1)^2\}.$$

Investigate and show that the products of the form ab with $a, b \in S$ are all distinct.

9. Plot the graph $\sin^2(\pi x) - x^2 \cos(\pi x) - x$ in the range $-3 \le x \le 3$. Then find (numerically) all roots of the equation $\sin^2(\pi x) - x^2 \cos(\pi x) - x = 0$ lying between -3 and 3.

10. Define the $n \times n$ matrix $A_n = (a_{ij})$ whose i, j−th entries are

$$a_{ij} = \begin{cases} 1 & i = j \\ i^2 + j^2 & i \ne j \end{cases}$$

Show that for $1 \le n \le 10$ the determinant of A_n is negative for odd values and positive for even values of n.

11. For a given n, find the number of odd elements of the form $\binom{n}{k}$, where $1 \le k \le n$.

12. Define $S(k, n) = \sum_{i=1}^{n} i^k$.

    a. Using the function $S(k, n)$ show that for any n, $(1 + 2 + \cdots n)^2 = (1^3 + 2^3 + \cdots n^3)$.

    b. Show that for any n,

$$\sum_{a=0}^{n} \frac{S(2, 3a+1)}{S(1, 3a+1)}$$

is always a square number.

13. For integers $2 \le n \le 200$, find all n such that n divides $(n-1)! + 1$. Show that there are 46 such n.

14. The formula $e_{41} = n^2 + n + 41$ produces prime numbers for $0 \le n \le 39$ but not for $n = 40$. Check that for no i between 1 and 10000 does the formula $e_i = n^2 + n + i$ produce prime numbers for a larger interval starting at $n = 0$

15. Show that

$$\frac{1}{\cos(0°)\cos(1°)} + \frac{1}{\cos(1°)\cos(2°)} + \cdots + \frac{1}{\cos(88°)\cos(89°)} = \frac{\cos(1°)}{\sin^2(1°)},$$

where $n°$ stands for n degrees (as opposed to radians).

16. Let A_n be the $n \times n$ matrix with (i, j)-th entry equal to x^{i-j-ij}. For $n = 3$ and 4 find exactly all values of x for which the determinant of A_n is zero. Find all the distinct values of x for which the determinant of A_6 is zero.

17. Consider $f(x, y) = \sin(x+y)\cos(x^2-y) - \sin(y)$ and generate separately the graphs of $\frac{\partial^2 f}{\partial x^2}$ and $\frac{\partial^2 f}{\partial x \partial y}$ over the rectangle $-\pi \le x \le \pi$ and $-\pi \le y \le \pi$. Find the maximum of the function $\frac{\partial^2 f}{\partial x \partial y}$ in this area.

18. Show that among the first 450 Fibonacci numbers, the number of odd Fibonacci numbers is twice the number of even ones.

19. Define the functions $f(k) = \sum_{n=1}^{k}(-1)^n \frac{t^n}{n!}$ and $g(k) = \sum_{n=1}^{k}(-1)^n \frac{n!}{t^n}$. Show that
$$2 - f(2)g(2) = \frac{2}{t} + \frac{t}{2}.$$

20. **a.** Plot the graph
$$f(x) = \cos^2(x) - e^{-\sum_{k=1}^{30} \frac{\cos(kx)}{1+k}} + e^{-\sum_{k=1}^{50} \frac{\sin(kx)}{1+k}}$$
for x ranging over the interval $[0, 2\pi]$. Then find (numerically) the root of the equation $f(x) = 0$ lying between 4 and 5.

b. Plot the solution of the equation $x^2y + xy^2 - x^4 - y^4 = 0$.

21. Find two positive integers x and y such that the equation $f(x, y) = 2xy^4 + x^2y^3 - 2x^3y^2 - y^5 - x^4y + 2y$ gives the 10th Fibonacci number. Draw the graph of $f(x, y)$ where $0 \le x \le 10$ and $0 \le y \le 20$. The graph should show that $f(x, y)$ acquires some positive values in this range.

22. Let r be a positive real number such that $\sqrt[4]{r} - \frac{1}{\sqrt[4]{r}} = 14$. Using *Mathematica*'s symbolic capabilities, show that $\sqrt[6]{r} + \frac{1}{\sqrt[6]{r}} = 6$.

23. Show that among the first 500 Fibonacci numbers

a. 18 of them are primes.

b. None of them is divisible by 350 and only one is divisible by 150. Find that number.

24. **a.** Show that the only primes p less than 1000 such that the remainder when 19^{p-1} is divided by p^2 is 1 are $\{3, 7, 13, 43, 137\}$.

b. For the numbers n between 1 and 100, we are interested in the number of primes between n and $2n$ (including n). Show that $n = 51$ is the only number such that the number of primes between n and $2n$, inclusive, is 11.

25. Write the function

$$f(k) = \sin(x) + x + \frac{x^3}{1 \times 2 \times 3} + \frac{x^5}{1 \times 2 \times 3 \times 4 \times 5} + \cdots + \frac{x^k}{1 \times 2 \times \cdots \times k}$$

(k is odd). Now find all the roots of the equation $x^3 - f[99]$ lying between 0 and 2.

26. Plot the graph of the "cowboy hat" equation

$$\sin(x^2 + y^2)e^{-x^2} + \cos(x^2 + y^2)$$

as both x and y ranges from -2 to 2. Now plot the graph twice more, in each case changing the viewpoint to show the graph when it is viewed from above and below. What is the maximum value that the function takes in this range?

27. Define a matrix A by

$$A = \begin{pmatrix} 1 + x + y & 2x + y + y^2 \\ x + x^2 + 2y & 1 + x^2 + y^2 \end{pmatrix},$$

and find its determinant. Use `ContourPlot[]` to show where, in the rectangle $-20 \le x, y \le 10$, this determinant is zero.

28. A number is perfect if it is equal to the sum of its proper divisors, e.g., $6 = 1 + 2 + 3$ but $18 \neq 1 + 2 + 3 + 6 + 9$. For k between 1 to 500, show that if $2^k - 1$ is prime then $2^{k-1}(2^k - 1)$ is perfect. (Hint, have a look at `Divisors`).

29. Consider the following property: If $n = p_1^{k_1} \cdots p_t^{k_t}$ is the decomposition of n into prime factors then $n = k_1 \times p_1 + k_2 \times p_2 + \cdots + k_t \times p_t$. Show that among the numbers from two to one million, 4 is the only non-prime number with this property.

30. Factorise the expression

$$(1 + x)^{30} + (1 - x)^{30}.$$

Your answer should be of the form

$$c f_1 f_2 f_3 f_4$$

where c is a constant, f_1 is a quadratic in x, f_2 is a polynomial of degree 4 in x, f_3 is of degree 8 and f_4 is of degree 16 in x. Find all the roots of the equations $f_1 = 0$, $f_2 = 0$ and $f_3 = 0$ exactly and find the roots of $f_4 = 0$

numerically. Show that all these roots are purely imaginary (i.e. of the form $0 + y\imath$ where y is a real number). Use *Mathematica* to find the minimum value that the original expression takes as x varies over the real numbers.

31. Plot a graph of the expression

$$x(2\pi - x) \sum_{n=1}^{50} \frac{\sin(nx)}{n}$$

for $0 \le x \le 2\pi$ using Plot's default settings. Point out the obvious inaccuracies in this graph. Make Plot produce a correct graph.

32. Define

$$f_n = \frac{x^{n+1} + y^{n+1}}{x^n + y^n}$$

for x and y real, and consider the expression

$$p_n = \frac{\partial^2 f_n}{\partial x^2} + \frac{\partial^2 f_n}{\partial y^2} - 2\frac{\partial^2 f_n}{\partial x \partial y}.$$

a) Find simple expressions for p_1, p_2 and p_3.

b) Show that p_1 never takes the value zero, no matter what real values x and y take.

c) Show also that p_2 takes the value zero on three straight lines in the plane, except at exceptional points where it is undefined. Give the equations of those three straight lines.

d) The expression p_3 also is zero on a number of straight lines (except at exceptional points where it is undefined). How many such lines are there and what are their equations?

33. Let A denote the 3×3 matrix

$$\begin{pmatrix} x + y & x^2 + y & x^3 + y \\ x + y^2 & x^2 + y^2 & x^3 + y^2 \\ x^2 + y^3 & x^2 + y^3 & x^3 + y^3 \end{pmatrix}$$

(yes, the fact that the bottom left-hand entry does not fit the pattern of the rest of the entries is intended!) and let B denote its inverse. Show that the sum of the entries in the first row of B is 0. What is the sum of the entries in each of the second and third rows? Also find the sum of the entries in each column of B.

34. For 32 of the positive integers n between 1 and 100 the number n^2+n+1 is prime. Which n are these? For how many of the positive integers n between 1 and 100,000 is n^2+n+1 prime?

35. Plot the points (x, y, z) in three dimensional space for which $x^2 + z^2 \leq 1$ and $y^2 + z^2 \leq 1$.

    Also produce a picture of that part of this set which remains when points for which $x^2 + y^2 < 1/4$ are removed.

36. Produce pictures of each of the following sets of points:

    a) Those (x, y) for which $x^4 + y^4 = 1$, $-1 \leq x \leq 1$ and $-1 \leq y \leq 1$.

    b) Those (x, y) for which $x^5 - 2xy + y^6 = 1$, $-2 \leq x \leq 2$ and $-2 \leq y \leq 2$.

    Superimpose these two images to see where the two curves cross. How many crossing points are there? Verify that all but one of these points lie on one or other of the axes and find their coordinates exactly. Use FindRoot[] to find the coordinates of the remaining crossing point.

37. If a and b are real numbers, show that the real part of

    $$\tan(a + bi)^2 + \tan(a - bi)^2 + \tan(-a + bi)^2 + \tan(-a - bi)^2$$

    is equal to

    $$-\frac{2(-2 + \cos(4a) + \cosh(4b))}{\big(\cos(2a) + \cosh(2b)\big)^2}.$$

    Find a similar expression for the imaginary part.

38. For 5 of the positive integers n between 1 and 100 the number $n^n + n + 1$ is prime. Which n are these? For how many of the positive integers n between 1 and 500 is $n^n + n + 1$ prime?

39. If a and b are real numbers, show that the real part of

    $$\big(\cos(a + bi) + \sin(a + bi)\big)^2$$

    is equal to

    $$1 + \sin(2a)\big(\cosh(b)^2 + \sinh(b)^2\big)$$

    and find a similar expression for its imaginary part.

40. Let $A = (a_{ij})$ denote the 4×4 matrix with

$$a_{ij} = x^i + x^j + x^{ij}.$$

What is the determinant of A? Find numerically those real values of x for which the determinant vanishes. If $x = \sqrt{2}$, what is the inverse of A?

41. Let $g(x) = \sin(x) + \cos(x)$. Plot graphs of $g(g(g(g(g(x)))))$ and $g(g(g(g(g(g(x))))))$ for x lying between 0 and π. There are four points where the graphs cross. Find, numerically, their (x, y) coordinates.

42. Let M be a general 2×2 matrix

$$\begin{pmatrix} a & b \\ c & d \end{pmatrix}$$

Show that if

$$M.M = \begin{pmatrix} a^2 & b^2 \\ c^2 & d^2 \end{pmatrix} \quad \text{and} \quad M.M.M = \begin{pmatrix} a^3 & b^3 \\ c^3 & d^3 \end{pmatrix}$$

then M must be of one of the following five types:

$$\begin{pmatrix} a & 0 \\ 0 & d \end{pmatrix} \quad \begin{pmatrix} a & 0 \\ a & 0 \end{pmatrix} \quad \begin{pmatrix} a & a \\ 0 & 0 \end{pmatrix} \quad \begin{pmatrix} 0 & d \\ 0 & d \end{pmatrix} \quad \begin{pmatrix} 0 & 0 \\ d & d \end{pmatrix}.$$

43. Let A_n be the $n \times n$ matrix with (i, j)-th entry equal to a^{i-j+ij}. For $n = 2, 4$ and 5 find exactly all values of a for which the determinant of A_n is zero. (For $n = 3$ the determinant is zero for all values of a.) Find numerically those values of a for which the determinant of A_6 is zero. Use `ListPlot[]` to show where these 22 solutions lie in the complex plane.

44. **(a).** Write a function to produce the series

$$g(k) = \frac{\sqrt{2}}{2} \frac{\sqrt{2 + \sqrt{3}}}{3} \frac{\sqrt{2 + \sqrt{3 + \sqrt{4}}}}{4} \cdots \frac{\sqrt{2 + \sqrt{3 + \cdots + \sqrt{k}}}}{k}.$$

Show that $g[100]$ is less than 10^{-100}.

(b). Write a function to produce the series

$$h(k) = \frac{\sqrt{2}}{2} \frac{\sqrt{2 + \sqrt{2}}}{2} \frac{\sqrt{2 + \sqrt{2 + \sqrt{2}}}}{2} \cdots \frac{\sqrt{2 + \sqrt{2 + \cdots + \sqrt{2}}}}{2}.$$

(The last factor has k copies of 2 in the numerator.) Show that the difference of $h[10]$ and $2/\pi$ is less than 10^{-5}.

17.2 Projects B

1. Consider the following construction. Start with a horizontal line of length 100 units. Fix two angles $\alpha = 62$ and $\beta = 60$ and a length $l = 40$. At the end of the line draw another horizontal line of length l. Then start to zigzag by drawing a sequence of lines of length $100, l, 100, l, 100, \cdots$, the anticlockwise angles these lines make with the horizontal being $\alpha, \beta, 2\alpha, 2\beta, 3\alpha, 3\beta, \cdots$ Write a program to repeat this construction 180 times. Produce the graph.

2. Consider a direct graph, consisting of vertices and edges such that each edge has a source and a range. See the following example (see also Chapter 16.1).

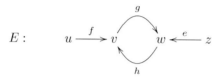

Here f is an edge with source u, denoted by $s(f)$, and range v, denoted by $r(f)$. A path is a sequence of edges $f_1 \cdots f_n$ such that the range of f_{i-1} is the source of f_i, i.e., $r(f_{i-1}) = s(f_i)$, $2 \le i \le n$. A path such as $f_1 \cdots f_n$, is called a *cycle* if $s(f_1) = r(f_n)$ and $s(f_i) \ne s(f_j)$, $i \ne j$. For example, gh is a cycle, whereas the path fg is not a cycle. A cycle $f_1 \cdots f_n$ has an *exit* if there is an edge e such that $e \ne f_i$, $1 \le i \le n$, and $s(e) = s(f_j)$ for some j.

For a graph E, let $n_{v,w}$ be the number of edges with source v and range w. Then the *adjacency matrix* of the graph E is $A_E = (n_{v,w})$. Usually one orders the vertices and then writes A_E based on this ordering.

Consider an $n \times n$ matrix with non-negative entires, and write a program to determine if the graph represented by this matrix has a cycle. If this is the case, then determine if it has a cycle with an exit. Furthermore, determine if all the cycles have exits.

3. Write a function copymatrix[] such that if A and B are matrices then copymatrix[A,B,col,row] makes a copy of B inside A by replacing A[[i+col,j+row]] by B[[i,j]]. You may assume that the copy of B fits inside A. Hence write a function bigmatrix[] such that if M1, M2, M3 and M4 are $n \times n$ matrices, then bigmatrix[M1,M2,M3,M4] is the $2n \times 2n$ matrix formed by arranging the entries in these four matrices thus:

$$\begin{pmatrix} M1 & M2 \\ M3 & M4 \end{pmatrix}.$$

Define a sequence of matrices A_n of size $2^n \times 2^n$ by defining

$$A_1 = \begin{pmatrix} 1 & -1 \\ 1 & 1 \end{pmatrix}$$

and (if $n \geq 1$)

$$A_{n+1} = \begin{pmatrix} A_n & -A_n \\ A_n & A_n \end{pmatrix}.$$

Find the determinant of A_n for $n = 1, 2, 3, 4, 5$.

Let $J = \begin{pmatrix} 1 & 0 \\ 0 & 1 \end{pmatrix}$ and $o = \begin{pmatrix} 0 & 0 \\ 0 & 0 \end{pmatrix}$.

Check the following identity for a 2×2 matrix A,

$$\begin{pmatrix} A & o \\ o & A^{-1} \end{pmatrix} = \begin{pmatrix} J & A \\ o & J \end{pmatrix} \begin{pmatrix} J & o \\ -A^{-1} & J \end{pmatrix} \begin{pmatrix} J & A \\ o & J \end{pmatrix} \begin{pmatrix} o & -J \\ J & 0 \end{pmatrix}.$$

18
Solutions to the Exercises

This chapter provides solutions to selected exercises.

Exercise 2.1

In order to get the correct result, one needs to group the items in the right order. Another approach is to use the *Mathematica* palettes and type the expression as given into the Front end.

```
Sqrt[64^(1/3)*(2^2 + (1/2)^2) - 1]
4
```

Exercise 2.2

We will enter the expression into *Mathematica* using Basic Math Assistant in Palettes. We will then try to use Simplify and, if that is not successful, we will try FullSimplify.

$$\left(\tfrac{1}{2} + \mathrm{Cos}\left[\tfrac{\pi}{20}\right]\right) \left(\tfrac{1}{2} + \mathrm{Cos}\left[\tfrac{3\pi}{20}\right]\right) \left(\tfrac{1}{2} + \mathrm{Cos}\left[\tfrac{9\pi}{20}\right]\right) \left(\tfrac{1}{2} + \mathrm{Cos}\left[\tfrac{27\pi}{20}\right]\right)$$

$$\left(\tfrac{1}{2} + \mathrm{Cos}\left[\tfrac{\pi}{20}\right]\right) \left(\tfrac{1}{2} + \mathrm{Cos}\left[\tfrac{3\pi}{20}\right]\right) \left(\tfrac{1}{2} + \mathrm{Sin}\left[\tfrac{\pi}{20}\right]\right) \left(\tfrac{1}{2} - \mathrm{Sin}\left[\tfrac{3\pi}{20}\right]\right)$$

$$\mathtt{Simplify}\left[\left(\tfrac{1}{2} + \mathrm{Cos}\left[\tfrac{\pi}{20}\right]\right) \left(\tfrac{1}{2} + \mathrm{Cos}\left[\tfrac{3\pi}{20}\right]\right) \left(\tfrac{1}{2} + \mathrm{Cos}\left[\tfrac{9\pi}{20}\right]\right) \left(\tfrac{1}{2} + \mathrm{Cos}\left[\tfrac{27\pi}{20}\right]\right)\right]$$

$$\left(\tfrac{1}{2} + \mathrm{Cos}\left[\tfrac{\pi}{20}\right]\right) \left(\tfrac{1}{2} + \mathrm{Cos}\left[\tfrac{3\pi}{20}\right]\right) \left(\tfrac{1}{2} + \mathrm{Sin}\left[\tfrac{\pi}{20}\right]\right) \left(\tfrac{1}{2} - \mathrm{Sin}\left[\tfrac{3\pi}{20}\right]\right)$$

$$\mathtt{FullSimplify}\left[\left(\tfrac{1}{2} + \mathrm{Cos}\left[\tfrac{\pi}{20}\right]\right) \left(\tfrac{1}{2} + \mathrm{Cos}\left[\tfrac{3\pi}{20}\right]\right) \left(\tfrac{1}{2} + \mathrm{Cos}\left[\tfrac{9\pi}{20}\right]\right) \left(\tfrac{1}{2} + \mathrm{Cos}\left[\tfrac{27\pi}{20}\right]\right)\right]$$

$$\tfrac{1}{16}$$

© Springer International Publishing Switzerland 2015
R. Hazrat, *Mathematica®: A Problem-Centered Approach*, Springer Undergraduate
Mathematics Series, DOI 10.1007/978-3-319-27585-7_18

Exercise 2.3

```
PrimeQ[123456789098765432111]
True
```

Exercise 2.4

We check this for $n = 5$.

```
n = 5;
LCM[1, 2, 3, 4, 5]
60

2^(5 - 1)
16
```

Exercise 2.5

One can enter 32! straight into *Mathematica* with confidence.

```
32!
263130836933693530167218012160000000
```

Clearly, the number ends with 7 zeros.

Exercise 2.6

```
Factor[(1 + x)^30 + (1 - x)^30]
2 (1 + x^2) (1 + 14 x^2 + x^4) (1 + 44 x^2 + 166 x^4 + 44 x^6 +
    x^8) (1 + 376 x^2 + 4380 x^4 + 15944 x^6 + 24134 x^8 +
    15944 x^10 + 4380 x^12 + 376 x^14 + x^16)
```

Exercise 2.7

```
Together[1/(1 + x) + 1/(1 + (1/(1 + x)))]
(3 + 3 x + x^2)/((1 + x) (2 + x))

Apart[%]
1 + 1/(1 + x) - 1/(2 + x)
```

Exercise 2.8

```
(1 + Sin[x] - Cos[x])/(1 + Sin[x] + Cos[x]) == Tan[x/2]
(1 - Cos[x] + Sin[x])/(1 + Cos[x] + Sin[x]) == Tan[x/2]

Simplify[(1 + Sin[x] - Cos[x])/(1 + Sin[x] + Cos[x]) ==
    Tan[x/2]]
True
```

Exercise 3.1

```
f[x_] := Sqrt[1 + x]

f[f[f[f[x]]]]
Sqrt[1 + Sqrt[1 + Sqrt[1 + Sqrt[1 + x]]]]
```

Exercise 4.1

```
Clear[f]

f[n_] := n^6 + 1091

RandomInteger[{1, 3095}]
590

Table[PrimeQ[f[RandomInteger[{1, 3095}]]], {6}]
{False, False, False, False, False, False}
```

Exercise 4.3

Among the first 450 Fibonacci numbers, the number of odd Fibonacci numbers is:

```
Length[Select[Fibonacci[Range[450]], OddQ]]
300
```

Thus the number of even Fibonacci numbers is 150, which is half of the number of odd ones.

Exercise 4.4 We first define A as a function which depends on m.

```
f[m_] := ((m + 3)^3 + 1)/(3 m)

fQ[m_] := IntegerQ[((m + 3)^3 + 1)/(3 m)]

Select[Range[500], fQ]
{2, 14}

f /@ %
{21, 117}
```

This shows that, for two integers m less than 500, $f(m)$ is an integer and in both cases they are odd numbers.

There are other ways to approach this problem using Count and Cases. These approaches will be discussed in Chapter 10.

```
Cases[f[Range[500]], _Integer]
{21, 117}

Count[f[Range[500]], _Integer]
2
```

Exercise 4.5

```
Length[Select[Range[20000], Divisible[#^2 + (# + 1)^2, 1997] &]]
20

Length[Select[Range[20000], Divisible[#^2 + (# + 1)^2, 2009] &]]
0
```

The following approach uses pattern matching, which will be discussed in Chapter 10.

```
Count[(Range[20000]^2 + (Range[20000] + 1)^2)/1997, _Integer]
20
```

Exercise 4.6

```
Select[Range[2, 200], Divisible[(# - 1)! + 1, #] &]
{2, 3, 5, 7, 11, 13, 17, 19, 23, 29, 31, 37, 41, 43, 47, 53, 59, 61,
67, 71, 73, 79, 83, 89, 97, 101, 103, 107, 109, 113, 127, 131, 137,
139, 149, 151, 157, 163, 167, 173, 179, 181, 191, 193, 197, 199}
```

```
Length[%]
46
```

Here is another way to write the code:

```
Length[Select[Range[2, 200], Mod[(# - 1)! + 1, #] == 0 &]]
46
```

Exercise 4.7

```
Select[Prime[Range[200]], Mod[19^(# - 1), #^2] == 1 &]
{3, 7, 13, 43, 137}
```

Exercise 4.8

```
re[n_] := FromDigits[Reverse[IntegerDigits[n]]]
```

```
re[12345]
54321
```

We need to look up to the 670-th prime number as

```
Prime[670]
5003
```

```
pless5000 = Prime[Range[670]];
```

```
Short[Select[pless5000, PrimeQ[re[#]] &]]
{2,3,5,7,11,13,17,31, <<151>>, 3803,3821,3851,
3889,3911,3917,3929}
```

```
Length[Select[pless5000, PrimeQ[re[#]] &]]
167
```

Exercise 4.9

```
Select[Range[1000], IntegerQ[Sqrt[#! + (# + 1)!]] &]
{4}
```

Yet another way to show there is only one n with the given property is by using the pattern matching of Chapter 10.

```
Count[Sqrt[Range[1000]! + (Range[1000] + 1)!], _Integer]
1
```

Or

```
Cases[Sqrt[Range[1000]! + (Range[1000] + 1)!], _Integer]
{12}
```

Exercise 4.10

```
Select[Range[10000], PrimeQ[#^6 + 1091] &, 5]
{3906, 4620, 5166, 5376, 5460}
```

Exercise 4.11

```
sr[n_] := Sort[IntegerDigits[n]]

cyclic[n_] := Length[Union[sr /@ (n Range[6])]] == 1

Select[Range[100000, 999999], cyclic]
{142857}

142857 * Range[6]
{142857, 285714, 428571, 571428, 714285, 857142}
```

Exercise 5.1

```
Select[Range[10, 99], Divisible[#, Plus @@ IntegerDigits[#]] &]
{10, 12, 18, 20, 21, 24, 27, 30, 36, 40, 42, 45, 48, 50, 54, 60,
63, 70, 72, 80, 81, 84, 90}

Length[Select[Range[10000, 99999],
  Divisible[#, Plus @@ IntegerDigits[#]] &]]
10334
```

Exercise 5.3

```
FactorInteger[2^4* 5^3* 7^3]
{{2, 4}, {5, 3}, {7, 3}}

Times @@@ FactorInteger[2^4* 5^3* 7^3]
{8, 15, 21}
```

As one can guess from the code, @@@ goes into the second level of the lists and
replaces the heads in that level.

```
Plus @@ (Times @@@ FactorInteger[2^4* 5^3* 7^3])
44

Select[Range[100], Plus @@ (Times @@@ FactorInteger[#]) == # &]
{1, 2, 3, 4, 5, 7, 11, 13, 17, 19, 23, 29, 31, 37, 41, 43, 47, 53,
59, 61, 67, 71, 73, 79, 83, 89, 97}
```

Exercise 5.4

```
Power @@ (x + y)
x^y

Plus @@@ (x^y + y^z)
x + 2 y + z
```

Exercise 5.5

This exercise again demonstrates that *Mathematica* is quite at ease when working with large numbers. First we get all the divisors of this large number. Then we add them together. Then we need to show that this number is a perfect number.

```
t = Divisors[60865556702383789896703717342431696226578307
3351885970528324860512791691264];
```

```
t1 = Plus @@ t
144740111546664524427946373126085988481573677491474835
89066354349131199152128
```

We have seen the following code in Problem 5.6, where we discussed perfect numbers.

```
Plus @@ Most[Divisors[t1]]
144740111546664524427946373126085988481573677491474835
89066354349131199152128
```

Exercise 5.6

We define a function to test if a number is perfect. Here we use the command Total to write this function. (Compare this with a code written in Problem 5.6).

```
?Total
Total[list] gives the total of elements in list.
```

```
perNumber[n_] := n == Total[Most[Divisors[n]]]
```

```
perNumber[6]
True
```

```
Select[Range[20], perNumber[2^(# - 1) (2^# - 1)] &]
{2, 3, 5, 7, 13, 17, 19}
```

Exercise 5.7

We first need to be able to produce the subsets of all the divisors. This can be done using Subsets.

```
A = {a, b, c}
{a, b, c}
```

```
Subsets[A]
{{}, {a}, {b}, {c}, {a, b}, {a, c}, {b, c}, {a, b, c}}
```

Next we need to exclude the empty set and the set of all the divisors from this list. This can be done at least in two ways:

```
Cases[Subsets[A], _?(# != {} && # != A &)]
{{a}, {b}, {c}, {a, b}, {a, c}, {b, c}}

Select[Subsets[A], 0 < Length[#] < Length[A] &]
{{a}, {b}, {c}, {a, b}, {a, c}, {b, c}}
```

We apply these to a specific number.

```
Divisors[122]
{1, 2, 61, 122}

A = Most[Divisors[122]]
{1, 2, 61}

Subsets[A]
{{}, {1}, {2}, {61}, {1, 2}, {1, 61}, {2, 61}, {1, 2, 61}}

Select[Subsets[A], 0 < Length[#] < Length[A] &]
{{1}, {2}, {61}, {1, 2}, {1, 61}, {2, 61}}
```

Next we need to consider the sum of these subsets. This can be done by changing the head of a list to a plus. However, since our list itself contains other lists as its members (the subsets), we need to use @@@ to change the head of the second layers.

```
Plus @@@ Select[Subsets[A], 0 < Length[#] < Length[A] &]
{1, 2, 61, 3, 62, 63}
```

All we need to do now is to check if our number is in this list. One option is to use FreeQ.

```
?FreeQ
FreeQ[expr,form] yields True if no subexpression in expr matches
form, and yields False otherwise

FreeQ[Plus @@@ Select[Subsets[A], 0 < Length[#] < Length[A] &] ,
122]
True
```

Putting all this together and designing a Do loop to run the code for $1 \leq n \leq 100$, we are done.

```
Do[With[{A = Most[Divisors[n]]},
  If[(Plus @@ A) > n &&
    FreeQ[Plus @@@ Select[Subsets[A], 0 < Length[#] < Length[A] &],
      n], Print[n]]],
  {n, 2, 100}]
```

70

Here is a slightly different approach.

```
a = Select[Range[100],
        Plus @@ Most[Divisors[#1]] > #1 & ]
```

```
{12, 18, 20, 24, 30, 36, 40, 42, 48, 54, 56, 60, 66, 70, 72, 78,
80, 84, 88, 90, 96, 100}
```

```
Select[a,
    !MemberQ[Apply[Plus, Subsets[Most[Divisors[#1]]]], 1], #1] & ]
    {70}
```

Can you write a faster code so that we would be able to run the code up to 1000?

Exercise 7.1

```
Clear[n, k]
```

```
Sum[k/(k^4 + k^2 + 1), {k, 1, n}]
(n + n^2)/(2 (1 + n + n^2))
```

Exercise 7.2

```
Clear[f, g, n]
```

```
f[k_] := Sum[(-1)^n t^n/n!, {n, 1, k}]
```

```
g[k_] := Sum[(-1)^n n!/t^n, {n, 1, k}]
```

```
2 - f[2] g[2] == 2/t + t/2
True
```

Exercise 7.3

```
f[k_] := Sin[x] + Sum[x^i/i!, {i, 1, k, 2}]
```

```
f[3]
x + x^3/6 + Sin[x]
```

Exercise 7.4

```
NSum[(-1)^n/(2 n + 1) Sum[1/(2 n + 4 k + 3), {k, 0, 2 n}],
{n, 0, Infinity}]
0.250902
```

```
N[(3 Pi/8 ) Log[(Sqrt[5] + 1)/2] - Pi/16 Log[5]]
0.250902
```

Exercise 7.5

```
t = Table[
    Sum[(-1)^i (1/(i + 1) + 1/(i + 2) + 1/(i + 3)), {i, 0, n,3}],
    {n, 10, 10000, 50}];

ListPlot[%]
```

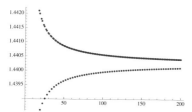

Exercise 7.6

```
k=1;Sum[Sum[Prime[k++], {j, i, 2 i - 1}]^(-1/2), {i, 1, 10}]
3/(2 Sqrt[2]) + (5 Sqrt[3])/44 + 1/(6 Sqrt[13]) + 1/Sqrt[23] +
1/( 3 Sqrt[30]) + 1/(2 Sqrt[33]) + 1/Sqrt[83] + 1/Sqrt[197]

ListPlot[Table[k=1;
    Sum[Sum[Prime[k++], {j, i, 2 i - 1}]^(-1/2), {i, 1, n}],
    {n, 1, 2000,  50}]]
```

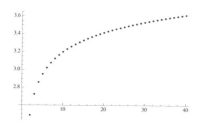

Exercise 7.7

```
Clear[f]

f[n_] := Product[Fibonacci[i] + x^i, {i, 1, n}]

f[3]
(1 + x) (1 + x^2) (2 + x^3)

Coefficient[f[23], x^4]
4487512294054883449227803198105834085114984960000
```

Exercise 8.1

We use two loops, a Do loop to change i and a While loop to check for what n the formula $n^2 + n + i$ is prime. That is *while* the formula $n^2 + n + i$ is prime, we add one to n and test the formula again: While[PrimeQ[n^2 + n + i]

For any n such that the above formula is prime, we collect them in the set A using AppendTo[A, n].

```
n = 0; A = {}; Do[
  While[PrimeQ[n^2 + n + i],
  AppendTo[A, n]; n++
  ];
  If[(A != {} && A != {0}) , Print[i, "  ", A]]; n = 0; A = {},
  {i, 1, 100, 2}
  ]
```

3   {0,1}

5   {0,1,2,3}

11   {0,1,2,3,4,5,6,7,8,9}

17   {0,1,2,3,4,5,6,7,8,9,10,11,12,13,14,15}

29   {0,1}

41   {0,1,2,3,4,5,6,7,8,9,10,11,12,13,14,15,16,17,18,19,20,21,22,
23,24,25,26,27,28,29,30,31,32,33,34,35,36,37,38,39}

59   {0,1}

71   {0,1}

We change the above code slightly to get the list of consecutive prime numbers for $n^2 + n + i$ where i is a prime number. For this we change the formula to n^2 + n + Prime[i] and let i run from 1 to 100 with step 2.

```
n = 0; A = {}; B = {}; Do[
  While[PrimeQ[n^2 + n + Prime[i]],
  AppendTo[A, n]; n++
  ]; AppendTo[B, {i, Length[A]}]; n = 0;
  A = {},
  {i, 1, 100, 2}
  ]; B
```

```
{{1, 1}, {3, 4}, {5, 10}, {7, 16}, {9, 1}, {11, 1}, {13, 40},
{15, 1}, {17, 2}, {19, 1}, {21, 1}, {23, 1}, {25, 1}, {27, 1},
{29, 1}, {31, 1}, {33, 2}, {35, 2}, {37, 1}, {39, 1}, {41, 2},
{43, 3}, {45, 2}, {47, 1}, {49, 4}, {51, 1}, {53, 1}, {55, 1},
{57, 2}, {59, 1}, {61, 1}, {63, 1}, {65, 1}, {67, 1}, {69, 5},
{71, 1}, {73, 1}, {75, 1}, {77, 1}, {79, 1}, {81, 2}, {83, 2},
{85, 1}, {87, 1}, {89, 3}, {91, 1}, {93, 1}, {95, 1}, {97, 1},
{99, 1}}
```

```
BarChart[Last /@ B]
```

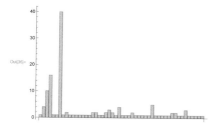

Exercise 8.2

```
n = t = 99999; i = 1;
While[MemberQ[IntegerDigits[t], 9],
i++;t=n*i]; Print[i, "    ", t]
```

```
11113    1111188888
```

Exercise 8.3

```
re[n_] := FromDigits[Reverse[IntegerDigits[n]]]
```

```
re[12345]
54321
```

```
Do[
 Do[
  If[IntegerQ[Sqrt[m^2 + n^2]] && m == re[n], Print[n, "    ", m ]],
  {m, n, 1000}],
 {n, 1, 1000}]
```

As we don't get an answer, there are no Pythagorean pairs smaller than
1000 which are reverses of each other.

Exercise 8.4

```
Input["Enter a prime number", p];
Do[
 Do[
  If[r^2 - q^2 == p^2, Print[r, "    ", q ]],
  {r, q + 1, 101}],
 {q, 1, 100}]
```

```
13   12
```

Exercise 8.5

```
f[x_] := 1/(1 + x)

Nest[f, x, 3]
1/(1 + 1/(1 + 1/(1 + x)))

NSolve[x == Nest[f, x, 10], x]
{{x -> -1.61803}, {x -> 0.618034}}
```

Exercise 8.6

```
f[n_] := Nest[Sqrt[2 + #] &, Sqrt[2], n]

f[3]
Sqrt[2 + Sqrt[2 + Sqrt[2 + Sqrt[2]]]]

f[4]
Sqrt[2 + Sqrt[2 + Sqrt[2 + Sqrt[2 + Sqrt[2]]]]]

t = N[Table[Product[f[n]/2, {n, 0, k}], {k, 1, 15}]]
```

```
{0.653281, 0.640729, 0.637644, 0.636876, 0.636684, 0.636636,
0.636624, 0.636621, 0.63662, 0.63662, 0.63662, 0.63662, 0.63662,
0.63662, 0.63662}

2./Pi
0.63662
```

Exercise 8.7

We use the code from Exercise 5.6 to find the sum of the proper divisors. Then we use NestList to repeatedly apply this function to a given number. If a number is social, it should be in the list generated by NestList.

```
NestList[Total[Most[Divisors[#]]] &, 1264460, 5]
{1264460, 1547860, 1727636, 1305184, 1264460, 1547860}
```

Since 1264460 is in this list, it is a social number. Here is a social number with a much longer cycle:

```
NestList[Total[Most[Divisors[#]]] &, 14316, 28]
{14316, 19116, 31704, 47616, 83328, 177792, 295488, 629072,
589786, 294896, 358336, 418904, 366556, 274924, 275444, 243760,
376736, 381028, 285778, 152990, 122410, 97946, 48976, 45946,
22976, 22744, 19916, 17716, 14316}
```

Exercise 8.8

```
x = {a, b, c, d, e, f}

Inner[Rule, x, RotateLeft[x], List]
{a -> b, b -> c, c -> d, d -> e, e -> f, f -> a}
```

Inner is in a way a generalisation of the command Thread. We could also have written

```
Thread[Rule[x, RotateLeft[x]]]
{a -> b, b -> c, c -> d, d -> e, e -> f, f -> a}
```

Exercise 8.9

```
karprekarRoutine[n_] :=
 Subtract @@
  FromDigits /@ (Sort[IntegerDigits[n], #] & /@ {Greater, Less})

NestList[karprekarRoutine, 5643, 10]
{5643, 3087, 8352, 6174, 6174, 6174, 6174, 6174, 6174,
 6174}

FixedPointList[karprekarRoutine, 7642]
{7642, 5175, 5994, 5355, 1998, 8082, 8532, 6174, 6174}

Length[FixedPointList[karprekarRoutine, #]] & /@
  Range[1000, 9999] // Max
 9
```

Exercise 9.1

```
(a + b)^n /. (x_ + y_)^z_ -> (x^z + y^z)
a^n + b^n
```

In general, $(a+b)^n$ is not $a^n + b^n$. However, in Ring theory, if the characteristic of a commutative ring is n, then $(a+b)^n = a^n + b^n$

Exercise 10.2

First, using pattern matching, we replace u^2 with its equivalent in terms of v. The following code does this systematically.

```
u^5 + u^4 + u^2 /. u^n_Integer -> x^Quotient[n, 2] u^Mod[n, 2]
```

We use this substitution to replace u with v. The rest is easy to follow from the codes.

```
(85 v^2 + 484 v - 313)^4 + (68 v^2 - 586 v +
   10)^4 + (2 u)^4 /.
u^n_Integer -> (22030 + 28849 v - 56158 v^2 +
    36941 v^3 - 31790 v^4)^Quotient[n, 2] u^Mod[n, 2]

(10 - 586 v + 68 v^2)^4 + (-313 + 484 v + 85 v^2)^4 +
16 (22030 + 28849 v - 56158 v^2 + 36941 v^3 - 31790 v^4)^2
```

```
Expand[(10 - 586 v + 68 v^2)^4 + (-313 + 484 v + 85 v^2)^4 +
16 (22030 + 28849 v - 56158 v^2 + 36941 v^3 - 31790 v^4)^2]

17363069361 - 39031031952 v + 101206498140 v^2 -
127484664912 v^3 + 167211563014 v^4 - 125377480368 v^5 +
97888478940 v^6 - 37127423088 v^7 + 16243247601 v^8

Expand[(357 v^2 - 204 v + 363)^4]

17363069361 - 39031031952 v + 101206498140 v^2 -
127484664912 v^3 + 167211563014 v^4 - 125377480368 v^5 +
97888478940 v^6 - 37127423088 v^7 + 16243247601 v^8

% == %%
True
```

Exercise 10.3

In Problem 4.5 a primitive version of this approach was used, i.e., using Factor. Here we use pattern matching to recognise when the polynomial can be decomposed into the product of other polynomials.

```
n = 5; While[! MatchQ[Factor[x^(n - 4) + 4 n], a_*b_],
n++]; n

16
```

Exercise 10.7

```
Select[DictionaryLookup[],
 MatchQ[Characters[#], {___, "r", "a", "t"}] &]

{"Ararat", "aristocrat", "autocrat", "baccarat", "brat",
"bureaucrat", "carat", "democrat", "Democrat", "Dixiecrat",
"drat", "frat", "Gujarat", "karat", "Marat", "Montserrat",
"Murat", "muskrat", "plutocrat", "prat", "rat", "Seurat", "sprat",
"Surat", "technocrat"}
```

Exercise 11.1

```
tt[x_, y_, z_] := Which[x == y, x, True, z]

tt[2, 3, 4]
4

tt[2, 2, 4]
2
```

Exercise 12.1

```
a[1] = 7;
a[n_] := a[n] = a[n - 1] + GCD[n, a[n - 1]]

Do[
 If[(t = a[n] - a[n - 1]) != 1, Print[n, " gives the prime ",
 t]], {n, 2, 500}]
```

```
5 gives the prime 5
6 gives the prime 3
11 gives the prime 11
12 gives the prime 3
23 gives the prime 23

24 gives the prime 3
47 gives the prime 47
48 gives the prime 3
50 gives the prime 5
51 gives the prime 3
101 gives the prime 101

102 gives the prime 3
105 gives the prime 7
110 gives the prime 11
111 gives the prime 3

117 gives the prime 13
233 gives the prime 233
234 gives the prime 3
467 gives the prime 467
```

Exercise 12.2

```
f[1] = 1; f[3] = 3;

f[n_?EvenQ] := f[n] = f[n/2]

f[n_?(Mod[#, 4] == 1 &)] := f[n] = 2 f[(n - 1)/2 + 1] -
 f[(n - 1)/4]

f[n_?(Mod[#, 4] == 3 &)] := f[n] = 3 f[(n - 3)/2 + 1] -
 2 f[(n - 3)/4]

Length[Select[Range[2015], f[#] == # &]]
```

```
93
```

Exercise 12.3

```
Clear[b]

b[1] := x
b[n_] := b[n] = b[n - 2] + x^n/n!

b[9]
x + x^3/6 + x^5/120 + x^7/5040 + x^9/362880
```

Exercise 12.5

```
f[m_, n_] := f[m, n] = (f[m - 1, n] + f[m, n - 1])

f[m_, n_] := Subscript[x, m] /; m == -n

phi[s_] := s /. Subscript[x, n_] -> Subscript[x, -n]

phi[f[2, 5]] == f[5, 2]
True
```

Exercise 13.2

A smart way to define this function is to use the command `Partition`.

```
?Partition[list,n] partitions list into non-overlapping
sublists of length n

d[n_] := Partition[Range[n^2], n]

Table[Det[d[n]], {n, 1, 10}]
{1, -2, 0, 0, 0, 0, 0, 0, 0, 0}
```

This shows, starting from $n > 2$, that the determinant is always 0.

If you insist on defining this matrix using `Array`, here is one way to do so:

```
d[n_] := Array[#2 + (#1 - 1)* n &, {n, n}]
```

Exercise 13.3

```
Clear[b, x]

b[1] = b[2] = 1; b[3] = 2;

b[n_] := b[n] = x /. (Flatten[Solve[Det[({
          {b[n - 3], b[n - 3] + b[n - 2], 1},
          {b[n - 1], b[n - 1] + x, 0},
          {0, 0, 1}
          })] == 1, x]])

b[4]
3

b[6]
11

b[123]
38705296961136956048990243108213402
```

Exercise 14.2

```
f[x_] := Sin[x] - Exp[-Sum[Cos[k x]/k, {k, 1, 30}]] +
    Exp[-Sum[Sin[k x]/k, {k, 1, 50}]]

Plot[f[x], {x, 0, 3 Pi}, PlotRange -> All, PlotPoints -> 50,
    MaxRecursion -> 2]
```

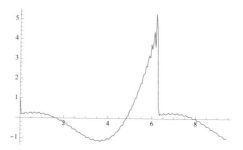

Figure 18.1 Exercise 14.2

```
NMaximize[f[x], {x, 0, 8}]
{4.47294, {x -> 6.0993}}
```

Although the maximum produced by *Mathematica* is about 4.47, a glance at the graph shows the maximum of this function is around 6. Let us draw the graph around the region where that maximum is attained and change the interval slightly.

```
Plot[f[x], {x, 5, 6.5}, PlotRange -> All, PlotPoints -> 50,
  MaxRecursion -> 2]

NMaximize[f[x], {x, 0, 7}]
{6.07758, {x -> 6.22194}}
```

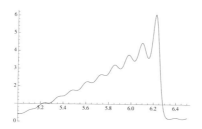

Figure 18.2 Exercise 14.2

Changing the interval to $[0, 7]$ produces the correct maximum. Why is this so? Investigate this.

Exercise 14.3

```
Plot[x (2 Pi - x) Sum[Sin[n x]/x, {n, 1, 50}], {x, 0, Pi},
  PlotRange -> All, PlotPoints -> 50, MaxRecursion -> 2]
```

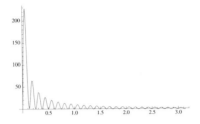

Figure 18.3 Exercise 14.3

Exercise 14.4

```
x[t_] := 4 Cos[-11 t/4] + 7 Cos[t]

y[t_] := 4 Sin[-11 t/4] + 7 Sin[t]

ParametricPlot[{x[t], y[t]}, {t, 0, 14 Pi}]
```

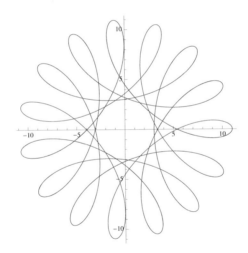

Figure 18.4 Exercise 14.4

Exercise 14.5

```
x[t_] := Cos[t] + 1/2 Cos[7 t] + 1/3 Sin[17 t]

y[t_] := Sin[t] + 1/2 Sin[7 t] + 1/3 Cos[17 t]

ParametricPlot[{x[t], y[t]}, {t, 0, 14 Pi}]
```

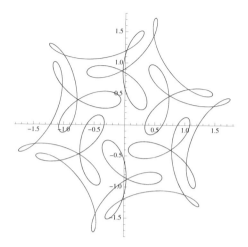

Figure 18.5 Exercise 14.5

Exercise 14.6

```
Plot[{2 Exp[-x^2], Cos[Sin[x] + Cos[x]]}, {x, -Pi, Pi}]
```

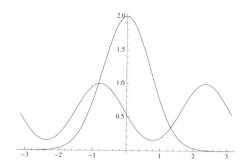

Figure 18.6 Exercise 14.6

Exercise 14.7

```
ContourPlot[ Abs[3 x^2 + x y^2 - 12] == Abs[x^2 - y^2 + 4],
{x, -10, 10}, {y, -10, 10}]
```

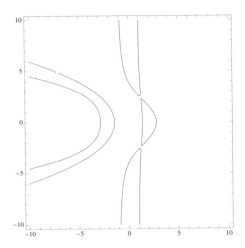

Figure 18.7 Exercise 14.7

Exercise 14.8

```
RegionPlot[Abs[x^2 + y] <= Abs[y^2 + x], {x, -5, 5}, {y, -5, 5}]
```

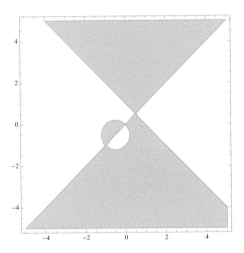

Figure 18.8 Exercise 14.8

Exercise 14.9

```
ContourPlot[4 (x^2 + y^2 - x)^3 - 27 (x^2 + y^2)^2 == 0,
{x, -4, 4}, {y, -4, 4}]
```

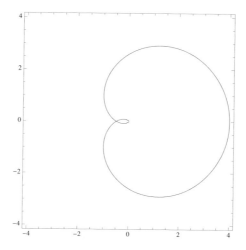

Figure 18.9   Exercise 14.9

Exercise 14.10

```
ContourPlot3D[x^2 y z + x^2 z^2 == y^3 z + y^3,
{x, -3, 3}, {y, -3, 3},   {z, -3, 3}, PlotPoints -> 50]
```

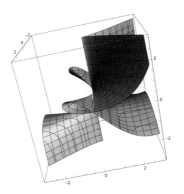

Figure 18.10   Exercise 14.10

Exercise 15.1

We define the function $f(x)$ in *Mathematica* and then first use NSolve to see if we are lucky enough to get all the roots of the equation.

```
f[x_] := x^3 - 3 x + 1

NSolve[f[f[x]] == 0, x]

{{x -> -2.09198}, {x -> -1.83201}, {x -> -1.63544},
{x -> -0.179284}, {x -> 0.221174}, {x -> 1.04599 - 0.531313 I},
{x -> 1.04599 + 0.531313 I}, {x -> 1.61084}, {x -> 1.81472}}
```

We get 7 real roots and 2 complex roots, a total of 9 roots. Since the degree of the equation $f(f(x))$ is 9, (by the fundamental theorem of algebra) these are all the roots of the equation.

Exercise 15.3

We first solve the system of equation for $r = 0$.

```
Solve[{x - 2 y == -1, -x + 3 y == 3}, {x, y}]
{{x -> 3, y -> 2}}
```

Next we solve the same system when $r = 0.2$.

```
Solve[{x - 2 y == -1, (-1 + 0.2) x + 3 y == 3}, {x, y}]
{{x -> 2.14286, y -> 1.57143}}
```

We will experiment with the error introduced in the coefficient, i.e., r, to see how the solutions will change. In order to do this, we define the solve function as a pure function and plug the errors in the place of r.

```
Solve[{x - 2 y == -1, (-1 + #) x + 3 y == 3}, {x, y}] & /@
  Range[0, 3, 0.2]

{{{x -> 3., y -> 2.}}, {{x -> 2.14286, y -> 1.57143}},
{{x -> 1.66667,  y -> 1.33333}}, {{x -> 1.36364, y -> 1.18182}},
{{x -> 1.15385,  y -> 1.07692}}, {{x -> 1., y -> 1.}},
{{x -> 0.88235, y -> 0.94117}}, {{x -> 0.78947, y -> 0.89473}},
{{x -> 0.714286, y -> 0.857143}}, {{x -> 0.652174, y -> 0.826087}},
{{x -> 0.6, y -> 0.8}}, {{x -> 0.555556, y -> 0.777778}},
{{x -> 0.517241, y -> 0.758621}}, {{x -> 0.483871, y -> 0.741935}},
{{x -> 0.454545, y -> 0.727273}}, {{x -> 0.428571, y -> 0.714286}}}

p = Flatten[
  Solve[{x - 2 y == -1, (-1 + #) x + 3 y == 3}, {x, y}] & /@
    Range[0, 5, 0.2], 1]

{x, y} /. p

{{3., 2.}, {2.14286, 1.57143}, {1.66667, 1.33333}, {1.36364,
  1.18182}, {1.15385, 1.07692}, {1., 1.}, {0.882353,
  0.941176}, {0.789474, 0.894737}, {0.714286, 0.857143}, {0.652174,
  0.826087}, {0.6, 0.8}, {0.555556, 0.777778}, {0.517241,
  0.758621}, {0.483871, 0.741935}, {0.454545, 0.727273}, {0.428571,
  0.714286}, {0.405405, 0.702703}, {0.384615, 0.692308}, {0.365854,
  0.682927}, {0.348837, 0.674419}, {0.333333, 0.666667}, {0.319149,
  0.659574}, {0.306122, 0.653061}, {0.294118, 0.647059}, {0.283019,
  0.641509}, {0.272727, 0.636364}}
```

We plot the solutions with the respect to the error introduced. From the graph below it is clear that the distance between the original solution is not linear.

```
Graphics[Point[{x, y} /. p]]
```

We define a function to find the solution for the error introduced into the system of the equations.

```
sol[r_] := {x, y} /.
    Flatten[Solve[{x - 2 y == -1, (-1 + r) x + 3 y == 3}, {x, y}] ]

sol[0]
{3, 2}

sol[1]
{1, 1}
```

Next we define a function to calculate the distance between the original solution and the solution after the error introduced into the system.

```
err[r_] := Module[{}, {s, t} = sol[0]; {a, b} = sol[r];
    Sqrt[Abs[a - s]^2 + Abs[b - t]^2]]

err[1]
Sqrt[5]
```

We plot the distance function err for $-5 \leq r \leq r$. From the graph below, it is clear that somewhere around -0.5 the distance tends to infinity.

```
Plot[err[r], {r, -5, 5}]
```

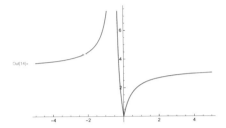

Note that in order to solve the equation we should calculate the inverse of the matrix.

```
Inverse[({{1, -2 }, {-1 + r, 3}})]
```

{{3/(1 + 2 r), 2/(1 + 2 r)}, {(1 - r)/(1 + 2 r), 1/(1 + 2 r)}}

Finally, we plot the lines, the distance function and the solutions all with the help of Manipulate.

```
Manipulate[GraphicsGrid[
  {{ContourPlot[{x - 2 y == -1, (-1 + r) x + 3 y == 3},
  {x, -15, 15}, {y, -15, 15}, PlotRange -> {{-20, 20}, {-20, 20}},
    Epilog -> {PointSize[Medium], Point[{3, 2}]}],
    Plot[err[g], {g, -10, r + 0.01},
    PlotRange -> {{-10, 10}, {0, 15}}]},
  {Graphics[{{PointSize[Large], Blue,
      Point[{3, 2}]}, {PointSize[Medium], Red, Point[sol[r]]}},
    PlotRange -> {{-10, 10}, {-10, 10}}], SpanFromLeft}},
  Frame -> All, FrameStyle -> Red,
  Background -> {{None, None}, {Yellow}}
  ],
  {{r, -10, "Error"}, -10, 10}]
```

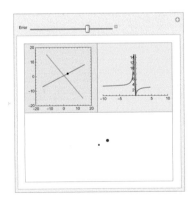

Further reading

Wolfram *Mathematica*® provides a collection of ready to use functions, and with its rules of programming it sets the stage like a chess board. Now it is up to you (and your imagination) how to combine these and make your move to attack the problem in hand. It always helps to look at different resources to get ideas of ways to combine the *Mathematica* functions.

There are excellent books written about *Mathematica*, for example Vardi [6], Wagon [7], Shaw–Tigg [5] and Gaylord, Kamin, Wellin [3, 8] to name a few. The reader is encouraged to have a look at them.

Wolfram demonstration projects `http://demonstrations.wolfram.com/` contains many interesting examples of how to use *Mathematica* in different disciplines.

Finally, the *Mathematica* Help and its virtual book is a treasure, dig it!

© Springer International Publishing Switzerland 2015
R. Hazrat, *Mathematica*® : *A Problem-Centered Approach*, Springer Undergraduate
Mathematics Series, DOI 10.1007/978-3-319-27585-7

Bibliography

[1] T. Andreescu, R. Gelca, Mathematical Olympiad Challenges, Birkhäuser, 2nd ed. 2009.

[2] R. Gaylord, *Mathematica* Programming Fundamentals, Lecture Notes, Available in MathSource

[3] R. Gaylord, S. Kamin, P. Wellin, An introduction to programming with *Mathematica*, Cambridge University Press, 2005.

[4] S. Rabinowitz, Index to Mathematical problems 1980–1984, Math pro Press. 1992.

[5] W. Shaw, J. Tigg, Applied *Mathematica*, Addison-Wesley Publishing, 1994.

[6] I. Vardi, Computational Recreations in *Mathematica*, Addison-Wesley Publishing, 1991.

[7] S. Wagon, *Mathematica* in Action, Springer-Verlag, 1999.

[8] P. Wellin, Programming with *Mathematica*: An Introduction, Cambridge University Press, 2013.

[9] E. Weisstein, MathWorld, `http://mathworld.wolfram.com/`.

© Springer International Publishing Switzerland 2015
R. Hazrat, *Mathematica®: A Problem-Centered Approach*, Springer Undergraduate
Mathematics Series, DOI 10.1007/978-3-319-27585-7

Index

© Springer International Publishing Switzerland 2015
R. Hazrat, *Mathematica®: A Problem-Centered Approach*, Springer Undergraduate
Mathematics Series, DOI 10.1007/978-3-319-27585-7

Printed in the United States
By Bookmasters